Phenomenology of Illness

The experience of illness is a universal and substantial part of human existence. Like death, illness raises important philosophical issues. But unlike death, illness, and in particular the experience of being ill, has received little philosophical attention. This may be because illness is often understood as a physiological process that falls within the domain of medical science, and is thus outside the purview of philosophy. In *Phenomenology of Illness* Havi Carel argues that the experience of illness has been wrongly neglected by philosophers and proposes to fill the lacuna. The book provides a distinctively philosophical account of illness. Using phenomenology, the philosophical method for first-person investigation, Carel explores how illness modifies the ill person's body, values, and world. The aim of *Phenomenology of Illness* is twofold: to contribute to the understanding of illness through the use of philosophy and to demonstrate the importance of illness for philosophy. Contra the philosophical tendency to resist thinking about illness, Carel proposes that illness is a philosophical tool. Through its pathologizing effect, illness distances the ill person from taken for granted routines and habits and reveals aspects of human existence that normally go unnoticed. *Phenomenology of Illness* develops a phenomenological framework for illness and a systematic understanding of illness as a philosophical tool.

Havi Carel is Professor of Philosophy at the University of Bristol, where she also teaches medical students. Her research examines the experience of illness and of receiving healthcare. She was recently awarded a Senior Investigator Award by the Wellcome Trust, for a five year project entitled 'Life of Breath' (with Prof Jane Macnaughton, Durham University). She has previously published on the embodied experience of illness, wellbeing within illness and patient-clinician communication in the *Lancet, BMJ, Journal of Medicine and Philosophy, Theoretical Medicine and Bioethics, Medicine, Healthcare and Philosophy,* and in edited collections. Havi is the author of *Illness* (2008, 2013), shortlisted for the Wellcome Trust Book Prize, and of *Life and Death in Freud and Heidegger* (2006). She is the co-editor of *Health, Illness and Disease* (2012) and of *What Philosophy Is* (2004).

This is a timely, important and beautifully written book. I fully concur with Havi Carel when she says, "It is not enough to see illness as an entity . . . that can be studied with the tools of science . . . it also has to be studied as a lived experience." Phenomenology could potentially revolutionalize how we understand illness and inform key improvements in how we diagnose, treat, and research it. This book would make an excellent addition to the core syllabus for students of medicine, nursing, and other professions as well as those of the humanities and social sciences.

Trish Greenhalgh, Professor of Primary Care Health Sciences,
University of Oxford

In this highly original and quite brilliant book, Havi Carel develops a sophisticated phenomenological approach to the study of illness, which does justice to the profundity, complexity, and diversity of illness experiences. The book offers a compelling and wide-ranging account of how philosophy can contribute to our understanding of illness and how the study of illness can enrich philosophy. More than that, it opens up a whole field for philosophical and interdisciplinary research, providing us with new ways of thinking about and responding to illness. This book will be essential reading for anyone seeking to better understand what it is to be ill.

Matthew Ratcliffe, Professor for Theoretical Philosophy,
University of Vienna

The experience of illness is a rich and multifaceted topic that has not gained sufficient attention, neither in medicine nor in philosophy. In this book Havi Carel remedies this lack by exploring the phenomenology of embodiment, vulnerability, and mortality. She develops a systematic and comprehensive account of how to think about being ill that will be immensely helpful to healthcare professionals and patients. And she shows how illness may be viewed as an unwelcome invitation to philosophize about the big topics of life and grow wiser in the process.

Professor Fredrik Svenaeus, Södertörn University

Havi Carel is amongst the most important voices developing a phenomenology of illness. Her work has completely transformed the way I think about illness and disease. *Phenomenology of Illness* is an important work that will become essential reading for philosophers of medicine, and will also be of great interest to healthcare professionals and all those touched by illness.

Dr Rachel Cooper, Senior Lecturer in Philosophy,
University of Lancaster

Phenomenology of Illness

Havi Carel

OXFORD
UNIVERSITY PRESS

OXFORD
UNIVERSITY PRESS

Great Clarendon Street, Oxford, OX2 6DP,
United Kingdom

Oxford University Press is a department of the University of Oxford.
It furthers the University's objective of excellence in research, scholarship,
and education by publishing worldwide. Oxford is a registered trade mark of
Oxford University Press in the UK and in certain other countries

First published 2016
First published in paperback 2018

Published in the United States of America by Oxford University Press
198 Madison Avenue, New York, NY 10016, United States of America

British Library Cataloguing in Publication Data
Data available

Library of Congress Cataloging in Publication Data
Data available

ISBN 978-0-19-966965-3 (Hbk.)
ISBN 978-0-19-882266-0 (Pbk.)

To Solomon, my brave and beautiful boy
To Joel, my funny valentine
To Samir, again and always

Contents

Acknowledgements

This book has emerged from a decade of writing, thinking, and speaking about illness, as well as living with it. Countless conversations, texts, events, emails, and images have contributed to the making of this book, and I am grateful to very many collaborators, interlocutors, friends, family, and colleagues.

I am grateful to the British Academy for awarding me a Mid-Career Fellowship in 2012–13, during which I wrote a first draft of this book. The Wellcome Trust awarded me a Senior Investigator award in 2014, and with the generous help of the Trust I completed work on the book. I thank Peter Momtchilhoff, commissioning editor for philosophy at Oxford University Press, for his support and guidance during the preparation of this book. Thanks also go to Eleanor Collins, Marilyn Inglis, Susan Frampton, and Banupriya Sivakaminathan, for their help with the production of the book. I am grateful to the two anonymous reviewers who made many helpful suggestions.

Work on the book started in earnest just as I joined the University of Bristol. I am grateful to my colleagues and collaborators here for fruitful discussions of illness and philosophy, and for the many ways in which friends and colleagues make life better. In particular I thank John Lee for pleasurable work together on the intercalated BA in medical humanities and for lovely French pastries. Joanna Burch-Brown suggested a useful restructuring of the book at an early stage. I thank Catherine Lamont-Robinson, an artist and arts and health practitioner and academic, and Louise Younie, a remarkable GP, for many shared conversations and joint work. I am grateful to Richard Pettigrew, the best colleague one could have, and to Giulia Terzian and Dagmar Wilhelm, for their friendship and support, often in the form of chocolate.

I am grateful to my previous colleagues at UWE, Bristol for their collegiality, friendship, and ideas. I thank Greg Tuck, who read draft chapters and commented on them in great detail. I thank Darian Meacham for comments on the book's material and for enjoyable work on a conference and edited volume on phenomenology and naturalism, from which I learned a great deal. I also thank Michael Lewis for fruitful discussions of Heidegger.

I am grateful to audiences at Durham, Bristol, Ryerson University, Canada, University College London, Exeter, Cumberland Lodge, UWE, Bristol, Sheffield, the British Society for Philosophy of Science Conference 2015, the Oxford Phenomenology Network, St Hughes, Oxford, Trinity College Dublin, Lancaster, Kings College London, Warwick, Reading, Swansea, Cambridge, Hull, Glasgow, Cardiff, Royal Holloway, Sussex, Newport, Bath, Hertfordshire, Mansfield College, Oxford, and the Jowett Society, Oxford, for inspiring thoughts and feedback on the material contained in this book.

I thank Matthew Ratcliffe for encouragement and support and for the many helpful conversations and useful comments on my work he provided over the years.

Ian James Kidd, with whom I co-authored work on epistemic injustice, has kindly given me permission to use our paper 'Epistemic injustice in healthcare: a philosophical analysis' in the book. Ian also read a draft of the book and provided many erudite suggestions. I am hugely grateful to him for allowing me to use our material in the book and for inspiring philosophical conversations.

My collaborator Jane Macnaughton on our Wellcome Trust project, the *Life of Breath* (<http://www.lifeofbreath.org>), has been a source of inspiration and support. I have learned a great deal from Jane and from our work together and am looking forward to continued collaboration on this and other projects. I am also grateful to Jess Farr-Cox, my black-belt project administrator, for being on the ball and knowing practically everything. Jess also proofread the book and prepared the index. I thank the entire *Life of Breath* team for their enthusiasm and ideas.

S. Kay Toombs's writings have been a profound influence on my work. Kay has also very kindly provided insightful and careful comments from the other side of the Atlantic. I hugely appreciate her time and thoughtfulness, as well as her friendship and support from afar.

I am grateful to the following publishers and journals for allowing me to use material in the book. 'Bodily doubt' was published in the *Journal of Consciousness Studies* 20(7–8): 178–97 (2013). 'Epistemic Injustice in Healthcare' was published as 'Epistemic injustice in healthcare: a philosophical analysis' (with I. J. Kidd), in *Medicine, Healthcare and Philosophy* 17(4): 529–40 (2014). 'The Philosophical Role of Illness' was published in *Metaphilosophy* 45(1): 20–40 (2014). Chapter 6 includes material from a previously published chapter: 'Ill, but Well: a Phenomenology of Wellbeing in Chronic Illness', in *Disability and The Good*

Human Life, J. Bickenbach, F. Felder, and B. Schmitz (eds.), Cambridge University Press (2013). And Chapter 2 contains material from 'Conspicuous, obtrusive, obstinate: a phenomenology of the ill body', in *Medicine and Society, New Perspectives in Continental Philosophy*, D. Meacham (ed.), Springer (2015) (with permission of Springer).

Finally, thank you, my family: my son Sol, who just turned eight, and like the sun shines bold and bright. My toddler Joel, whose quirky humour and baby softness are a daily reminder of why it is so good to perceive.

Thanks to my parents, whose unstinting support and many, many hours of childcare enabled this book to be written. I am especially grateful to my mother, Cynthia Carel, for her help and solicitude in all matters mothering and grandmothering, and for listening without judging.

I thank my wonderful sister, Sari Carel, who created the image on the cover.

And finally, thank you, Samir, for being the wisest and strongest person I know. I'm glad we still agree on what really matters.

Introduction

A process in the eye forewarns
The bones of blindness; and the womb
Drives in a death as life leaks out.

Dylan Thomas, 'A Process in the Weather of the Heart'

The experience of illness is a universal and substantial part of human existence. Like death, illness raises important philosophical issues. But unlike death, illness, and in particular the experience of being ill, has received little philosophical attention. This may be because illness is often understood as a physiological process that falls within the domain of medical science, and is thus outside the purview of philosophy. This book suggests that the experience of illness has been wrongly neglected by philosophers in general, philosophers of science and medicine in particular, and by biomedical theory and practice. This philosophical and biomedical neglect stands in stark contrast to the intense interest in this experience from the general public and from ill people and those who care for and about them.

I suggest that it is necessary to supplement a naturalistic account of disease (I use the term 'disease' to denote physiological dysfunction) with a philosophical study of the experience of illness (how the disease is experienced) itself. This approach aims to study illness without viewing it exclusively as a subject of scientific investigation. It is not enough to see illness as an entity in the world that can be studied with the tools of science. In order to fully understand illness it also has to be studied as a lived experience. To study the lived experience of illness is necessarily to explore its existential, ethical, and social dimensions.

Hence I propose to use a phenomenological approach for this study. Phenomenology—the philosophical method for studying lived experience—is not a variant form of scientific enquiry, but a method for examining pre-reflective, subjective human experience as it is lived prior to its theorization by science. Of course there can be a mutually illuminating dialogue between the two modes of study, as can be seen in recent work that crosses or explores the boundaries between phenomenology and cognitive science (see for example Petitot et al. 1999; Wheeler 2013; Carel and Meacham 2013; Gallagher and Zahavi 2008). But in what follows I focus on a philosophical phenomenological enquiry that can mutually inform and interact with scientific work while remaining independent of it.

This book is a comprehensive philosophical exploration of the experience of illness. The aim of the book is twofold: to contribute to the understanding of illness through the use of philosophy, and to demonstrate the importance of illness for philosophy. This bilateral approach lies at the heart of the book. It argues that a philosophical analysis is essential to developing a full understanding of illness, and complements work in medical anthropology, sociology of health and illness, health psychology, and qualitative health research.

The book also argues that illness is a rich and underexplored area for philosophy as a discipline, as it challenges central philosophical concepts in moral and political philosophy such as fairness, autonomy, and agency, but also in metaphysics, for example, by changing the experience of space and time. Illness does this by bringing normal processes and practices into sharp relief, juxtaposing them with cases of pathology and dysfunction. The book advances both the claim that philosophy is necessary for a full understanding of illness and that the study of illness is integral to a philosophical investigation of human existence (see Chapter 9).

Before progressing any further, let us pause and ask: what is illness, the subject of this book? By illness, I refer to serious, chronic, and life-changing ill health, as opposed to a cold or a bout of tonsillitis. The emphasis of this book is on the more profound phenomenological changes associated with serious illness, where the onset of illness is not followed by complete recovery within a short period of time. I suggest that illness experiences are heterogeneous and need to be studied in their particularity. However, I also claim that we can identify changes in the global structure of experience that apply to many, or even all, illnesses. Hence, it is the dual role of a phenomenological approach both to attend to the individual and idiosyncratic experience of illness as well as to discern systematic changes to the structure of experience brought about by illness.

It is important to note at the outset that illness is distinguished from disease, and an illness experience encompasses those phenomenological changes that can be directly or indirectly attributed to the effects of the disease process. However, some illnesses, for example, some kinds of mental disorder, may not involve disease (physiological dysfunction) at all. So are those conditions still to be characterized as illnesses, even if they are not tied to the presence of disease? In other words, what characteristics unite all and only illness experiences? I examine the question in detail in Chapter 1. The emphasis in this work is on those experiences that are indeed associated with disease, but I note that some experiences are sufficiently similar in at least some respects, to fall under the same phenomenological account, regardless of whether or not we choose to extend the term 'illness' to such cases.

Illness is a complex human puzzle. The more one reflects on it, the more pieces it seems to have and the more interconnected they appear. This book carries out the philosophical work of analysing this puzzle, drawing out such connections as well as articulating general insights and identifying blind spots in order to provide a comprehensive account of illness. It has grown out of a decade of thinking and writing about illness, as well as living with it. It has also grown out of talking about illness with philosophers, patients, health professionals, artists, novelists, medical educators, and people with ill relatives and friends. I have learned much from all of them. However, I have also come to think that without a comprehensive, systematic—and distinctively philosophical—account of illness we will continue to walk blindly through its complex topography, with no conceptual map to guide us. This book is intended to provide such a map, with the hope of tracing the terrain of illness, so as to bring this philosophically uncharted territory into clearer view.

Illness is a breathtakingly intense experience. It unsettles, and sometimes shatters, the most fundamental values and beliefs we hold. It is physically and emotionally draining. It can be physically and psychologically debilitating. Illness requires serious effort and continuous work to adapt practically to its limitations and to adjust psychologically to the pain, restricted horizons, and frustration it brings. It forces the ill person and those around her to confront mortality at its most direct and bare manifestation. In all of these ways illness requires labour, attention, and a conscious and sustained effort.

But, as this book argues, illness is also existentially and intellectually demanding—and potentially rewarding—in ways hitherto largely over-looked. Illness can challenge our most fundamental beliefs, expectations, and values and this accords it a distinct and important philosophical role. (Whether and how this role is recognized by ill people and by philosophy is discussed in the final chapter of this book.) For example, the belief that a longer life is better than a shorter life is one such belief that comes under intense scrutiny when one is given a poor prognosis. Perhaps a shorter but more meaningful life is equally valuable, or even better than a longer life? In what ways does the value of life depend on its duration? These are some of the philosophical questions that may arise in response to illness. I return to these questions in Chapter 7.

Here is another example of a philosophical theme that arises from illness. Many ill people report a change in their perception of space and time (Ratcliffe 2012a; Carel 2012). Places that were 'just round the corner', things that were 'easy to carry', and events that were vaguely conceived as 'a long way off' (such as death), change their function and as a result also their meaning. For example, stairs that once led some-where are now an obstacle for a paraplegic. Not only the contents of experience but also its structure and normative fit change in illness; this has obvious relevance to philosophical areas such as phenomenology, philosophy of mind, and philosophy of perception.

In the most general terms we can say that illness changes how the ill person experiences the world and how she inhabits it. Such a dramatic and overarching change to experience deserves, or often simply demands, philosophical attention. A common reaction to a diagnosis of an illness is a sense of meaninglessness and despair; such an event challenges the ill person to reflect on her life and search for ways of regaining meaning. As such, illness is one path leading into philosophical reflection by calling us to question our understanding of the world, fundamental beliefs, habits, and expectations. Illness often distances the ill person from her previous life and constitutes a violent invitation—or, again, *demand*—to philosophize. I examine this process in detail in Chapter 9.

What kind of invitation to philosophize illness is, and what philo-sophical work can arise from this invitation, are two major concerns of this book, which have so far garnered little attention from philosophers. In the final chapter I offer an account of the philosophical productiveness of illness, but also of its violent mode. Of course the invitation to reflect on one's life issued by illness can be met with resistance. Many people, both healthy and ill, deny, flee, or otherwise resist thinking about vulnerability, morbidity, and mortality. This reaction stems from the difficult nature of illness: unwelcome and demanding, it extends its bony fingers to grab, restrict, and sometimes destroy all that we hold dear: freedom, motility, agency, action, possibility, and the openness of the future. Perhaps philosophers' neglect of illness is one sort of such denial, or an attempted falsification (in the Nietzschean sense) of the realities of life through wilful, although often unconscious, blindness to its uglier faces.

Contra the philosophical tendency to resist thinking about illness, I propose that illness is a philosophical tool. Through its pathologizing effect illness distances the ill person from taken-for-granted routines and habits and thus reveals aspects of human existence that normally go unnoticed. For example, in Chapter 4, I claim that we have a tacit sense of bodily certainty that only comes to our attention when it is disrupted and replaced by bodily doubt. As such, illness is a useful philosophical tool for shedding light on the structure and meaning of both normal and pathological human experience.

Developing a systematic understanding of illness as a philosophical tool is another task the book sets out to achieve. It seeks to clarify and elaborate the philosophical value and usefulness of illness and to explore the ways in which philosophy may make use of illness. This, too, is something philosophers have not, with few exceptions (e.g. MacIntyre 1999), engaged in previously. But if we turn to sciences such as neurology or developmental psychology, we find that illness, or pathology more generally, has many established uses. For example, the study of the effects of brain injury can help us understand the normal function of a brain region (Gallagher 2005). The study of the results of lack of stimulation and nurturing in infancy can lead to insight about the contribution of such input to normal child development (Gerhardt 2004). Viewing illness as a subcategory of pathology and using it in an analogous way to understand the normal structure of human experience has much to offer philosophy.

What we have, therefore, is a bilateral flow from philosophy to illness and back. Philosophy, and in particular phenomenology, can be used to understand the experience of illness, and illness can be used to shed new light on central areas of philosophy such as ethics, political philosophy, and the study of human experience. The idea that the illumination is mutual is a central theme of this book. The book aims primarily to illuminate illness through the use of philosophical concepts, frameworks, and methods. But it also suggests that illness ought to be studied seriously by philosophers because it has a distinctive contribution to make to philosophical work in various areas of the discipline.

The philosophical claims advanced in this book will be illuminated by personal examples, quotations, and testimonials of ill people and of those who care for them, or are otherwise touched by illness. These will not serve merely as illustrations, vignettes, or anecdotes. A primary claim of this book is that the personal and anecdotal are essential to this kind of philosophical work and are indeed what motivates it. The personal suffering and personal growth which are respectively afflicted and afforded by illness cannot be abstracted from the particular context of an individual life and the idiosyncratic way in which illness is lived by a particular person.

The offer to take first-person reports and stories seriously is a challenge to some understandings of philosophy as an abstract and general reflection, or indeed as the formulation of universal rules to aid our understanding (e.g. logic) or guide our behaviour (e.g. ethics). Of course different types of philosophical work require different levels of abstraction. I am not suggesting that all of philosophy ought to proceed in the existential-phenomenological method proposed here. However, the practice of philosophy and the drive to philosophize are rooted in subjective, personal experience and can be characterized as an attempt to distinguish the idiosyncratic from the shared aspects of such experience.

Whenever we abstract, we abstract from a prior concrete experiential totality: the world as lived, or 'being in the world'. The purpose of abstraction is to understand that world and then return to it with new sensibilities. The source of philosophical insight is and remains subjective experience even if philosophy's ultimate aim is to understand its subject matter in a more objective or intersubjective manner. Phenomenology—or at least the kind of existential phenomenology employed here—is a process of improving our understanding of both general and concrete aspects of human experience.

Some have argued that the appeal to emotions and anecdote is an illegitimate philosophical move. Michael Sayeau, for example, has argued that such an appeal introduces an extra-philosophical dimension that should not be admitted into philosophy's realm (2009). In contrast, I suggest that emotion and anecdote are fundamental building blocks of human experience and should therefore be accorded a place within philosophical analysis, although this place needs to be carefully thought out and rigorously examined. What triggers an emotional, indeed an aesthetic, response to a first-person account of illness can be morally and existentially edifying and indeed yield epistemic rewards in philosophically relevant ways (Burley 2011; Nussbaum 1990; Kidd 2012). I provide such an account in Chapter 5, which utilizes the theoretical phenomenological framework developed in earlier chapters.

Structure and Method of the Book

This book sets out to examine illness philosophically, using a phenomenological approach. This seems particularly apt for this task for three reasons: first, phenomenology provides a framework that exposes the untenability of mind/body dualism and emphasizes the centrality and importance of the body to understanding human experience. Second, phenomenology is the study of lived experience, and as such is a natural method to use in order to study the variety of illness experiences. Third, phenomenology offers a non-prescriptive approach to illness. As such it provides a non-judgemental open framework that aims to free its practitioners from conceptual restrictions. Edmund Husserl (1988) famously advocated phenomenology as a 'presuppositionless' form of philosophical enquiry. Many suggest that such a demand can be, at best, a regulative ideal for phenomenology because it is unachievable in practice. But it remains an aspiration of phenomenological enquiry: it seeks to make explicit and suspend previously unacknowledged assumptions wherever possible, thus opening up new possibilities for understanding.

Using a phenomenological approach, different domains of illness are examined in the book. The first half of the book (Chapters 1 through 5) develops a general framework for a phenomenology of illness. Chapter 1 explains how phenomenology can help us understand the experience of illness, why phenomenology is the most apt approach to understanding this experience, and what aspects of human existence are particularly salient to this experience. The chapter begins by contrasting illness and disease and discussing the problematic tendency dominant in our culture of reducing the former to the latter. It then turns to mental and somatic illness and outlines their shared aspects. The chapter also explains the rationale behind a methodology focusing on first-person reports and sets out the possibilities, both theoretical and practical, for developing such an approach.

Chapter 2 offers a phenomenological framework through which to study illness. The framework developed in this chapter consists of five interconnected themes that offer a comprehensive and nuanced account of the experience of illness. The first theme is S. Kay Toombs's analysis of the shared features of illness as a series of five losses: loss of wholeness, certainty, control, freedom, and loss of the familiar world. The second theme is the distinction between the objective body and the body as lived, which Husserl and Merleau-Ponty put forward as a fundamental view of the body. This chapter demonstrates how this distinction is useful for understanding the difference between disease and illness (presented in Chapter 1), as well as for explaining communication problems in the clinic.

The third theme is Sartre's three orders of the body: the body as objective and as subjective, but also the order of intersubjectivity, which is the body as I experience it as reflected in the experience of others. I use Sartre's account to expound Husserl and Merleau-Ponty's view presented in the previous section. The third order of the body contains diverse experiences that make us conscious of how one's body is perceived by others and how this affects one's own understanding of their body. The fourth theme examines the claim that the healthy body is transparent, or absent, and illness is the loss of this transparency. I discuss the limitations of this view and ask whether the healthy body really is transparent, offering examples of lost transparency within health. Finally, using Heidegger's tool analysis I suggest that illness is a breakdown of 'bodily tools'. I argue that although the body is not a tool in the Heideggerian sense, his taxonomy of tool breakdown is analogous to, and hence useful for, an account of illness.

These five phenomenological analyses cumulatively provide us with a general framework for a phenomenology of illness. Chapter 3 uses this framework to describe the body in illness, positing illness as a unique way of being in the world. Using the framework developed in the previous chapter, the chapter examines how the lived body changes in illness, as well as how the physical and social worlds of the ill person are modified as a result of illness. The chapter opens with a fleshing out of

Toombs's account of illness as a series of losses. It then examines the change of spatial experiences in illness, and how, because of that, shared meanings and concepts change as well.

The chapter then turns to the social architecture of illness, and how illness modifies one's social relations in both positive and negative ways. The chapter concludes by examining how illness might modify Heidegger's account of human existence from 'being able to be' (*Seinkönnen*) to 'being unable to be'. Because options are closed down and possibilities rescinded, the being of the ill person has to be reconceived as 'inability to be' or 'partial ability to be'. I supplement Heidegger's account by broadening the range of 'abilities to be', using illness as a central example and justification for the modification of Heidegger's account.

Chapter 4 focuses on a particular aspect of illness experience, namely the experience of bodily doubt. This chapter suggests that what I call bodily certainty and its absence, bodily doubt, are core features of existential feelings, as characterized by Matthew Ratcliffe. I describe bodily doubt as a loss of continuity, loss of transparency of the body, and loss of faith in one's body. I suggest that bodily certainty is an instance of Hume's general claim about our animal nature: we trust that our bodies will continue to function as they do now. This certainty is not epistemically justifiable but is impossible to relinquish. The experience of bodily doubt makes this tacit certainty explicit and enables us to study it. The chapter then argues that the exploration of pathology is useful for highlighting tacit aspects of experience that otherwise go unnoticed. The end of the chapter places bodily doubt in broader context, suggesting that it provides a philosophical tool for studying human experience.

The final chapter of this first half of the book, Chapter 5, applies the framework developed thus far to respiratory illness. The chapter analyses respiratory illness and its main symptom, breathlessness, using the phenomenological account developed here. The experience of breathlessness is acutely distressing, and yet it is also invisible and difficult to both measure and describe. It is invisible because respiratory patients are often housebound or suffer from mobility problems, thus remaining hidden to the public as well as to clinicians. It is also invisible because the experience of breathlessness can seem innocuous to the external observer. However, the experience of breathlessness is debilitating, frightening, and ill-understood. The chapter uses the phenomenological tools developed in earlier chapters to outline a phenomenology of breath-lessness, and to explore the discrepancy between the subjective experience of breathlessness and the objective measurement of lung function.

The second half of the book examines how illness might affect our relationship to central philosophical themes: happiness, justice, embodi-ment, time, and death. Chapter 6 examines the relationship between ill health and well-being, basing its philosophical claims on empirical data demonstrating a lack of correlation between the two, dubbed 'the dis-ability paradox'. The chapter poses a question: how is well-being possible within the confines of poor health, an uncertain prognosis, and limita-tion of one's freedom? I suggest that two enigmas of health arise from this question. First, why don't ill people become more markedly unhappy as a result of their illness? Second, given that we have robust evidence showing that well-being is not impacted by ill health, why do we fear illness, and see it as an evil, when we are healthy?

I explain the discrepancy by looking carefully at what takes place when 'insiders' and 'outsiders' view illness and disability. I suggest some explan-ations for the remarkable resilience people show in the face of illness (and other adversity). This resilience stands in tension with the fact that illness is usually seen as a disaster that strikes an individual. But despite this view of illness as a tragedy, empirical and qualitative evidence shows that illness can also be an opportunity and a challenge. I suggest that the reason that we normally observe only the negative aspects of illness, and hence consider it a serious misfortune, is that one's ability to conceive of an ill life that is also happy is limited by one's 'outsider's perspective' on illness. Hence the importance of detailed phenomenological descriptions and first-person accounts of illness; they help communicate a thick account of illness that only intimate acquaintance with it can provide.

Chapter 7 provides an analysis of a concept crucial to any account of illness: death. I centre this analysis on Martin Heidegger's notion of life as a constant movement towards death: a 'being towards death', as he calls it. The chapter engages with Heidegger's death analysis, providing a new interpretation of the concept. I suggest that the views put forward by Hubert Dreyfus and William Blattner, namely that death is the condition of 'being unable to be anything' or of 'finitude of possibility', need augmenting. I amend their view by adding a second meaning to Heidegger's concept of death, that of temporal finitude. I also criticize Heidegger's view that authentically facing death demands individuation. Instead I suggest that we need a more relational understanding of death and illness, which would provide a more positive framework for understanding human finitude and vulnerability.

Chapter 8 turns to the notion of epistemic injustice, developed by Miranda Fricker (2007): a wrong done to someone specifically in their capacity as knower. In joint work with Ian James Kidd, we apply the notion of epistemic injustice to the case of illness and claim that ill people are particularly vulnerable to several kinds of epistemic injustice. We argue that ill people can suffer two kinds of epistemic injustice: testimonial and hermeneutical. In the case of illness, testimonial injustice can be suffered through the presumptive attribution of such characteristics as cognitive unreliability and emotional instability that downgrade the credibility of ill people's testimonies. Ill persons can also suffer hermeneutical injustice because many aspects of the experience of illness are difficult to understand and communicate and this is often owing to gaps in collective hermeneutical resources.

We then argue that epistemic injustice arises in part owing to the epistemic privilege enjoyed by the practitioners and institutions of contemporary healthcare services—the former owing to their training, expertise, and third-person psychology, and the latter owing to their implicit privileging of certain styles of articulating and evidencing testimonies in ways that marginalize ill persons. We end by suggesting that a phenomenological approach (e.g. the use of a phenomenological toolkit (Carel 2012)) may be part of an effort to ameliorate epistemic injustice.

So far the book uses philosophical ideas, methods, and concepts to develop an account of the experience of illness. The final chapter of the book, Chapter 9, travels in the opposite direction, asking not what philosophy can do for illness, but rather what illness can do for philosophy. The chapter articulates the philosophical role of illness. It briefly surveys the philosophical role accorded to illness in the history of philosophy and explains why illness merits such a role. It suggests that illness modifies, and thus sheds light on, normal experience, revealing its ordinary and therefore overlooked structure. Illness also provides an opportunity for reflection by performing a kind of suspension (*epoché*) of previously held beliefs, including tacit beliefs. These characteristics warrant a philosophical role for illness.

While the performance of most philosophical procedures is volitional and theoretical, illness is uninvited and threatening, throwing the ill person into anxiety and uncertainty. As such it can be viewed as a radical philosophical motivation that can profoundly alter our outlook. I suggest that illness can change how we philosophize: illness may shape philosophical methods and concerns and change one's sense of salience and conception of philosophy.

The book ends with George Canguilhem's definition of disease as 'a new way of life for the organism', the creation of new norms that govern the relationship of the diseased organism to its environment (1991, p. 84). The richness of the experience of illness and the understanding of health and illness as distinctly normative, attest to the fact that illness both requires and merits further philosophical exploration.

Two further comments are in order before we begin. The first is a useful distinction, offered by Ian James Kidd, between *pathophobic* and *pathophilic* attitudes towards illness (personal communication). Pathophobic attitudes are characterized by fearing illness and wishing to avoid it at all costs. These attitudes are common throughout the Western world and, in particular, promoted by the view that biomedical science and technology can resolve the problem of illness. But because such resolution is (as yet) impossible, these views often turn to denial of illness, and to rejection of the illness experience as potentially valuable and worthy of study.

Pathophilic attitudes are the opposite: they view illness as potentially edifying, positive, purifying, and instructive. Such attitudes can be found in the premodern world, where illness was unavoidable and thus best tackled with acceptance and preparation. We can find such views in the Stoics (see Chapter 9), some passages in Nietzsche (2004), and in Kidd (2012). On these views, illness is potentially edifying and instructive not only in a personal sense but also in a general philosophical sense.

However, both views can still hold that we can stand to gain philosophically from illness. On the pathophilic view, illness is a source for growth and edification. On the pathophobic view, we can still think that we might learn from bodily aberration, even if we abhor it. For example, David Hume (1975) writes: 'Thus the beauty of our person, of itself, and by its very appearance, gives pleasure, as well as pride; and its deformity, pain as well as humility.' The deformity of one's person, unwelcome though it may be, might still provide a valuable lesson in humility.

I suggest that we need to find a middle ground between the two positions. We ought to be prudentially pathophobic, but also to acknowledge that the experience of illness is surprisingly rich and diverse and can sometimes yield unexpected positive results. Although we should not court illness nor welcome it, we can turn our attention to these positive results, and look at the factors that can support our efforts to cultivate those. For example, resilience, gratitude, and intimacy are known to be three elements that are important for being able to cope well with illness (and other adverse life events) but are also themselves cultivated and further honed by adverse experiences (Gilbert 2006; Haidt 2006). We should pay more attention to experiences of growth in illness, and learn how to live well and cope well with illness. Reflective coping, of the kind I developed in my book, *Illness*, is a useful way to absorb the negative impacts of illness, but keep open the ability to experience wellness within illness (see Chapter 6).

Finally, I would like to suggest an observation about the ethics of phenomenology. In this book I try to heed the phenomenologists' call for attention to implicit structures of experience. This attention may reveal insights about our conditioned embodied existence. But this also means that being oblivious to the kinds of embodied certainty and freedom that we possess makes it difficult to empathize with radically different forms of embodiment or be open to the thought that there is something worthwhile in studying the minutiae of daily existence (for a different defence of this point see Dorfman 2014).

This, I think, is an ethics that is integral to the phenomenological method: a humbling recognition that our thought, experience, and activity fundamentally presuppose a way of being in the world, tacit bodily certainty, a sense of reality, and other features that are often taken for granted or seen as philosophically trivial, but in the absence of which life becomes profoundly difficult. That phenomenological sensitivity can translate into an ethical sensibility: it may call on us to develop an openness to others grounded in a robustly critical self-reflective study of oneself (compare Ratcliffe 2012b). Phenomenology is not a form of self-indulgence, but a mode of reflection that enables us not only to understand ourselves better, but also to direct our attention towards others in thoughtful empathy. I hope that this book will convince the reader of the importance of understanding our ways of living with illness through such a self-reflective study, but also that it opens new avenues for exploring how such self-understanding can cultivate an understanding of others who may be radically different.

1

Why Use Phenomenology to Study Illness?

This opening chapter suggests that phenomenology is apt for the philosophical study of illness. It is worth reminding the reader here again that the term illness is used within this work to denote serious and life-altering conditions, rather than mild ones, such as the common cold. This chapter begins by contrasting illness and disease and discussing the problematic tendency dominant in some medical quarters to reduce the former to the latter. It also discusses mental and somatic illnesses and outlines their shared aspects. The chapter then explains the choice of a methodology focusing on first-person reports and explicates the possibilities, both theoretical and practical, for developing such an approach to illness. It outlines the advantages of a phenomenology of illness, as uniquely suited to a philosophical analysis of one's experience of illness.

The chapter suggests that illness (as opposed to disease) is a complete transformation of one's life. As such we must enlist philosophical analysis in order to fully appreciate the existential transformation illness brings about. This transformation cannot be accounted for as merely physical or mental (in the case of psychiatric disorder) dysfunction. Rather, there is a need for a view of personhood as embodied, situated, and enactive, in order to explain how local changes to the ill person's body and capacities modify her existence globally. Although they did not write explicitly about illness (with the exception of some passages in Sartre) Husserl, Merleau-Ponty, Heidegger, and Sartre offer a phenomenological framework for understanding human experience in general. I adapt this framework and use it to develop a philosophical account of illness.

Illness is a breakdown of meaning in the ill person's life. Because of the disruption of habits, expectations, and abilities, meaning structures are destabilized and in extreme cases the overall coherence of one's life is destroyed; this is perhaps especially so in the case of mental illness, to which I return in §1.1. One's body, and the fundamental sense of embodied normalcy in which habits and values are rooted, is also disrupted. This disruption is often devastating but it is also an opportunity to examine the ill person's life. It can be used to explore an individual life, its meaning, goals, and values and how best to modify them in response to illness.[1]

Illness can also be used to explore how meaning and intelligibility depend on consistent patterns of embodiment. When these patterns are disrupted, meaning is affected. It is the second, more general sense that is investigated in the book. However, as was pointed out in the Introduction, the philosophical exploration in this book is grounded in particular experiences that beset ill people. These experiences are not simply to be mined and then discarded in favour of a generalization. They remain at the heart of the phenomenological analysis and the movement from subjective, idiosyncratic experience to more general philosophical analysis and back is intrinsic to this book.

Finally, this chapter suggests that illness is a 'deep phenomenon', an encounter with which reveals the lack of autonomy of rational subjectivity, and as such exposes the limits of this subjectivity. Another example of a deep phenomenon is art; Merleau-Ponty writes about Cezanne's paintings as showing us a world in which subjectivity recognizes its limited capacity to organize and structure experience (Merleau-Ponty 1964a). This experience of vulnerability, but also illumination, is fundamental to the experience of illness. It is this philosophical potential of illness that the book investigates.

1.1 Disease vs Illness

Modern medicine is characterized by a focus on biological dysfunction, its causes, and treatment, and by increasing reliance on medical technology for prevention, diagnosis, and treatment (Bishop 2011). The medical world view dominant in contemporary Western societies is scientific and relies heavily on understanding discrete mechanisms and

---

[1] The patient toolkit described in Chapter 8, this volume, aims at enabling this process.

functions in a detailed, if piecemeal, fashion.[2] It is also highly, and increasingly, specialized. Modern medicine has been extremely success-ful in, for example, understanding and controlling the causes of infec-tious disease.

Breaking the physiological body down to subsystems and functions and studying these discretely has yielded incredible results in under-standing cell proliferation processes (e.g. cancer) and understanding how body systems and organs function and how to intervene in useful ways when function is replaced by sub- or dysfunction. Modern medicine has made great advances in the last 150 years or so, in understanding what causes disease and what may ameliorate or prevent it. This brief descrip-tion is, of course, sketchy and there is much debate amongst historians and sociologists of medicine on many of its points. But it will suffice here as a general characterization of contemporary clinical medicine.[3]

In the modern-day institutions and healthcare arrangements that developed based on this understanding, the vast majority of medical practice, and especially acute and internal medicine, has remained firmly grounded in the understanding of disease as biological dysfunction and as the medically salient factor. Other areas of medicine are of their nature much more linked to the subjective, the personal, and the social. Paedi-atrics, geriatrics, palliative care, general practice, and psychiatry are notable exceptions to this medical model, or its later reincarnation, the biopsychosocial model, but are sadly perceived by some medical students and professionals as less prestigious areas of specialization.

The idea that the causes of disease are pathogens or human behaviour and that disease can be exhaustively described in third-person language has led many health professionals and kinds of medical and healthcare training to focus on disease. Other dimensions have been neglected, as we can see in many domains from hospital design and architecture, to the well-documented communication problems between healthcare staff and patients, to low patient satisfaction and service design (Korsch 1968, 1969; Coulter and Ellins 2006; compare Bishop 2011). These factors are

---

[2] I do not intend to give a detailed account of this world view here, since what drives our everyday tacit understanding of medicine and of illness is not a nuanced account, but an inherited set of assumptions (for a critical discussion see Goldacre 2009).

[3] For more on modern medicine and its achievements, see Porter (1999). For a critical view see Wootton (2006) and Bishop (2011).

not medically unimportant, but are often overlooked by institutions, policymakers, and practices that privilege disease over illness.

Illness is the experience of disease, the 'what it is like' qualitative dimension as it is experienced and made meaningful by the ill person. This includes the experience of one's symptoms and bodily changes, but also the experience of receiving healthcare and experiencing social attitudes towards illness and disability, pain, grappling with one's mortality, and negotiating what may become a hostile world. To take an example: if someone has cancer that is at a very early stage, and has no symptoms and no knowledge of the cancer, then that person is diseased, but is not ill (yet): they have no experience of illness. Conversely (and more contentiously), if one is mildly depressed and cannot pursue everyday activities but has no corresponding brain lesion, then one is ill but not diseased: there is no known physiological disease process. Disease is to illness what our physical body is to our body as it is lived and experienced by us.[4]

This book is about the experience of illness, not the facts of disease. Disease falls within the domain of empirical science, while illness has traditionally been understood as lending itself to analysis via disciplines such as psychology, sociology, anthropology, and law, and more recently via the medical humanities. I suggest that illness can—and should—be studied by philosophy as well, and moreover, that bilateral connections run from illness to philosophy and back.

The first claim about illness I wish to make is that viewed from a subjective or everyday perspective, illness is the most important element of the disease–illness coupling. We care about physiological dysfunction primarily when it causes us pain or discomfort, or prevents us from doing certain things, although of course merely knowing about a disease status or risk is enough to cause us stress and anxiety. Our contact with disease is through our experience of it, be that direct (experiencing symptoms) or indirect (learning about one's genetic risk factors). Although I focus on the disruption of first-person experience in the first part of the book, such disruption due to a disease process is not

---

[4] Of course mental illness poses a special challenge to this view, because it may be that some mental illnesses are caused by psychological dysfunction that arises only at a functionally defined psychological level, not in brain dysfunction. But it may also be that medical knowledge is currently unable to identify the disease, but will be able to do so in the future.

always the route through which we come to care about a disease. However, I suggest that this disruption is phenomenologically central because it allows us to examine the changed embodiment, uncanniness, bodily alienation, and bodily doubt that characterize it.

The second claim is that rather than seeing illness as secondary to disease, we ought to view it as a primary set of phenomena (Toombs 1988). If we add to these two claims a further claim—that illness is a fundamental experience in almost everyone's life—we can appreciate the importance of understanding how illness impacts upon one's life, how it changes experience, and how it shapes the life of the ill person, ultimately forming a 'complete way of being' (Merleau-Ponty 2012, p. 110).

Before we move on to phenomenology, we need to address the special issues that mental disorder raises with respect to the disease/illness distinction. I said earlier that disease is the physiological process, while illness is the experience of that process. But what about a case in which there is no known physiological process, as is the case with some mental disorders? Can we have illness without disease? And would this be a good way of characterizing some mental disorders (e.g. mild depression)? Psychiatric illness presents special difficulties in this regard. We are not yet able to say what the disease process is in many psychiatric disorders. However, within psychiatry there is now an increasing focus on molecular biology and genetics, and psychiatry as a whole is largely committed to the view that mental disorder is caused by abnormalities of the nervous system (at least in part, as many diseases are caused by multiple factors), or put more simply, that 'diseases have bodily sites' (Murphy 2006, p. 107 and ch. 4). One way of solving the problem would be to say that although at present we do not have sufficient knowledge of disease mechanisms in psychiatric disorder, it is likely that a physiological basis will eventually be found.

However, as Dominic Murphy (2006) notes, the problem is not merely an empirical issue of lack of scientific knowledge that will become available in the future. He writes, 'the current literature on mental illness lacks a coherent concept of the mental and a satisfactory account of disorder . . . ' (p. 6). It may be that the notion of mental disorder does not lend itself to the disease/illness distinction because the science of psychiatry is still in its infancy. But it may be that a process different to somatic illness causes mental disorder and that no amount of empirical investigation will uncover a disease process of the kind we have in diabetes or asthma. This complex issue will not be further addressed here, since this book focuses primarily on somatic illness, but I suggest a few observations.

First, some mental disorders do fit the disease/illness model, as they do have a clear physiological disease process underlying the illness symptoms. Dementia, a recognized psychiatric disorder classified as such by the *Diagnostic and Statistical Manual of Mental Disorders* (5th ed.) (DSM-5), is one such example. Dementia is caused by death of brain cells. This means that at least some mental disorders, for example neurodegenerative diseases, could be analysed using the disease/illness distinction in a way similar to somatic disorders.

Second, as excellent work in the field has demonstrated over the last fifty years, mental disorders lend themselves well to phenomenological analysis (see for example Jaspers 1997; Ratcliffe 2008a; Stanghellini 2004). In fact, there is currently no way, other than asking people to describe their experiences, to know about many psychiatric symptoms such as delusions, abnormal ideation, hearing voices, or visual hallucinations. Of course the behaviour of the person experiencing the symptoms often discloses the symptoms; but they may disclose no information in their behaviour (e.g. catatonic patients). The reliance on first-person reports is therefore important, and certainly more marked, in the diagnosis and treatment of some psychiatric conditions.

Third, phenomenology has played a role in the history of psychiatry that has not been paralleled in somatic medicine. So for our purposes here, we can rely on this body of literature in claiming the usefulness of first-person reports in psychiatry. For these reasons we should recognize that even if the disease process of some mental disorders is unknown or non-existent, the illness experience of mental disorder is as robust, important, and worthy of phenomenological analysis as its somatic counterpart. With these reflections in mind, let us now turn to this book's method: phenomenology.

1.2 What is Phenomenology?

Phenomenology is a philosophical tradition dating back to the early years of the twentieth century. Within the tradition there are different views and emphases, but most generally, phenomenology is a philosophical approach that focuses on phenomena (what we perceive and experience) rather than on the reality of things (what there is). It focuses on the experiences of thinking, perceiving, and coming into contact with the world: how phenomena appear to consciousness (Moran 2000, p. 1). Phenomenology examines the encounter between consciousness and the world, and views the latter as inherently human-dependent; as can be seen from its name, it is the science (*logos*) of relating consciousness to *phenomena* (appearances), rather than to *pragmata* (things as they are).

Phenomenology is primarily a descriptive philosophical method aimed at providing a philosophical description of consciousness and its engagement with the world. As such, phenomenology aims to be a practice rather than a system (Moran 2000, p. 4). As a practice, it has been used in a range of disciplines such as sociology, film studies, anthropology, nursing, musicology, and others. It can be used to describe particular experiences, for example, the experience of viewing a painting, as in Heidegger's analysis of Van Gogh's 1886 painting *Peasant Shoes* (Heidegger 1993, pp. 158–61). It can be used to describe how something appears from a certain point of view, given a certain environment, as in Merleau-Ponty's analysis of Cezanne's paintings (1964, pp. 9–25). Or it can be used to analyse the experience of listening to a melody, as discussed later in this chapter (Husserl 1990).

Phenomenology is used to attend to various aspects of our experience, providing a method for discerning and describing human experience. What phenomenologists aspire to discern and describe are the implicit structures of experience—that things do not simply appear to us as baldly *there*, but appear in a particular way, for example, as enticing, repellent, out of place, useful, and so on. How they appear as endowed with meaning is what phenomenology aims to provide an account of. Phenomenology is particularly useful not only for analysing discrete units of input (e.g. a spoken sentence) but in understanding the particular background against which the input is perceived and interpreted (e.g. a background of sexism providing a particular context to the sentence, compare MacKinnon 1993).

Phenomenology understands perceptual experience as embedded in a particular culture and as having a particular meaning based on the concepts and values of that culture. But it is not merely an anthropological method. Phenomenology is a distinctly *philosophical* method, as it investigates the conditions of possibility for having a particular experience; it is a transcendental method of inquiry, rather than an empirical one (Gallagher and Zahavi 2008, pp. 132–7). However, while phenomenology is normally described as a transcendental mode of inquiry, as we shall see, the boundaries between the transcendental and the empirical can become blurred. Some phenomenologists, such as Merleau-Ponty (2012) downplay the significance of the transcendental nature of phenomenology. For the purposes of describing and understanding the experience of illness, the more modest view is sufficient, according to which it is enough to consider the general features of illness without insisting on their transcendental nature.[5]

The main aim of phenomenology is to study perception, cognition, and other aspects of mental life in a non-empirical manner. That is, phenomenology does not seek causal explanations for empirical phenomena. Phenomenology is not a branch of empirical psychology and it does not ask, or attempt to answer, causal questions about specific mechanisms that give rise to perception, thought, consciousness, or other aspects of experience. However, phenomenology has an important relationship with empirical psychology, neuroscience, cognitive science, and other domains that study human consciousness in an empirical manner.

How does one study consciousness in a non-empirical manner? Here phenomenological approaches diverge. Edmund Husserl (1988), the founder of phenomenology, suggested that phenomenology was a transcendental science, that is, the study of the conditions of possibility of consciousness. Transcendental phenomenology does not posit phenomenal data as empirical, objectively real, or absolute, but rather as transcendental, i.e. as constrained by the conditions of possibility of consciousness. Its transcendental, rather than empirical, data is what differentiates phenomenology from experimental psychology. While experimental psychology generates empirical data about different mental acts, phenomenology generates transcendental data on the conditions of possibility of mental acts. As Kant has shown, space and time are such conditions of possibility, without which experience would be impossible (Kant 1999, A32/B48; Gardner 1999, pp. 75ff.).

---

[5] See Toombs (1988) for a transcendental analysis of illness, although she later rejected this view.

Husserl is a Kantian thinker in that he sees phenomena, what appears to us, as fundamentally constituted by consciousness. Phenomenal nature, for Kant, what was knowable, was a correlate of the activities of transcendental consciousness, the structure of which was knowable a priori. The laws of nature to which the sensible world conforms are in this sense knowable insofar as they are the laws of the mind according to which the world is constructed. But this limits cognition to the 'boundaries of possible experience', leaving 'the thing in itself as indeed real *per se*, but as not known by us'.[6] Husserl's project was to try and discover the rules that guide transcendental consciousness.

Phenomenology describes the mental activity taking place in different acts of consciousness, such as perceiving, thinking, imagining, and so on. An example may clarify the particular kind of descriptive philosophy that is phenomenology. In lecture notes collated in *On the Phenomenology of the Consciousness of Internal Time* (1990 [1966] *Husserliana* X), Husserl addresses the question: how must consciousness be structured in order to enable us to hear a melody? He suggests that consciousness requires the temporal structures of retention and protention in order to hear a melody. Retention enables us to retain the memory of the notes that have just been played, against which we hear the present tone. Protention extends into the future, presenting us with a sense of anticipation of what the coming notes might be. The frustration or fulfilment of this expectation determines, in part, the melody we hear.

A 'tone-now', he says, passes over into retention: it is held in present consciousness as a 'tone-having-been'. It is against the background of this tone that the next 'tone-now' is heard. And the connection between the two creates an experience of hearing a melody, rather than a succession of discrete tones. Similarly for protention—a present tone is heard against our expectation of what it may be. The interplay between past, present (constantly turning into past), and future (the tone about to come) creates a cohesive experience of a melody, which is a temporal entity.

---

[6] 'For we are brought to the conclusion that we can never transcend the limits of possible experience, though this is precisely what this science [metaphysics] is concerned above all else to achieve. This situation yields, however, just the very experiment by which, indirectly, we are enabled to prove the truth of this first estimate of our *a priori* knowledge of reason, namely that such knowledge has only to do with appearances, and must leave the thing in itself as indeed real *per se*, but as not known by us' (Kant 1999, Bxix–xx).

Husserl writes: 'The tone begins and "it" steadily continues. The tone-now changes into a tone-having-been; the *impressional* consciousness, constantly flowing, passes over into ever new *retentional* consciousness' (1991, p. 31). The conclusion Husserl draws from this example is that consciousness requires retention and protention in order to hear a melody. This investigation is not empirical in nature, but transcendental—examining the conditions of possibility for hearing a melody.

Other phenomenologists had different views of how phenomenology ought to proceed and what its metaphysical commitments are. Heidegger's dense prose focuses on the question of the meaning of being (*Die Seinsfrage*), asking what does it mean to be, to exist? In order to answer this question Heidegger articulates a description of the fundamental structures of the human being which he calls *Dasein* (1962). Dasein means 'existence' but Heidegger breaks the word down to its constituents. *Da-sein*, literally 'there-being' or 'being there', is the term he uses to denote the human being as already thrown into a world (1962).

Heidegger's analysis of Dasein contains transcendental components but is also heavily influenced by the cultural and social writers of his day, and is more concerned with the meaning of human existence than the study of acts of consciousness. The French phenomenologist Merleau-Ponty, despite his great interest in empirical science and psychology of his time, remained committed to the Husserlian view. However, Merleau-Ponty was not a transcendental thinker, and was committed to the 'third way' he developed, between empiricism and 'intellectualism' (rationalism), which enabled him to forge a new way of understanding the human being as a 'body-subject' (see §1.3 of this chapter).

Some thinkers, such as Sartre, held yet other views, and were influenced by other thinkers (e.g. Hegel and Marx).[7] The main distinction I wish to draw here is between transcendental phenomenology (e.g. the early Husserl) and existential phenomenology, concerned with pre-reflective everyday life, human subjective experience, and existential themes such as freedom and authenticity, found in the work of the late Husserl, as well as Heidegger, Merleau-Ponty, and Sartre. I will mainly be

---

[7] For a comprehensive historical account see Moran (2000).

using the existential phenomenologists here, but will frequently refer back to Husserl as well, whose ideas were foundational.[8]

How should we proceed in the study of a phenomenology of illness, given such a range of views? I suggest that for this task specifically, phenomenology may be considered metaphysically modest: it focuses on the data available to human consciousness, while bracketing metaphysical debates and ontological commitments to the existence of external objects. The point of phenomenology is to make clear that metaphysical theorizing can proceed only on the basis of implicit structures of experience that can only be described phenomenologically. In this sense phenomenology is a 'fundamental science'. It is this return to experience itself that this book utilizes in studying the experience of illness.

What sort of phenomenological approach would be best suited to the description of the experience of illness? Regardless of the various differences between the phenomenologists mentioned earlier, we can draw out common themes which are significant to the understanding the phenomenology of illness. In what follows I develop an approach that accounts for the body's central role in human life and acknowledges the primacy of perception. Such an approach is found in the work of Husserl and in the writings of Merleau-Ponty. Both thinkers developed a phenomenology of embodiment; Husserl in Division II of *Ideas II* and Merleau-Ponty (2012) in his 1945 work *Phenomenology of Perception*. It is these two texts and their understanding of the human being as a 'body-subject' to which I now turn.

1.3 Phenomenology and Embodiment

Husserl viewed phenomenology as the 'presuppositionless science of consciousness' (Moran 2000, p. 126). This starting point takes nothing for granted and suspends our usual understanding of the phenomena that appear to us as real things located in a mind-independent space. This suspension of underlying metaphysical beliefs plays a major role in phenomenological method and supports the metaphysical modesty discussed earlier in this chapter. Husserl wished for phenomenological

---

[8] This brief overview will suffice for the purposes of this book, but there are a great number of introductions to and overview of the phenomenological tradition that the reader is invited to consult for further study (Cerbone; Moran; Baldwin; and others).

investigation to embrace this modesty as its starting point: 'I have thereby chosen to begin in absolute poverty, with an absolute lack of knowledge' he says in the opening pages of the *Cartesian Meditations* (1988, p. 2), echoing Descartes's own quest, in his *Meditations* to devote himself 'sincerely and without reservation' to the general demolition of his opinions and search for a foundation for these beliefs (Descartes 1988, p. 76). Husserl's suspension of belief involves not assuming any philosophical or scientific theory, and rejecting even deductive reasoning, mathematics, and empirical science, in order to focus on describing what is given directly in experience. This became a radical form of self-questioning, involving questioning our natural intuitions about consciousness and the world (Moran 2000, p. 126). 'Phenomenology must return to what is directly given in exactly the manner in which it is given'; in order to achieve that we must attend to our intuitions in an appropriate manner (2000, p. 126.). Our mental acts are self-given in an absolute manner and their study is the 'pure science of phenomenology'.

Husserl wanted to break away from actuality in order to study 'imaginative free variation', a method for uncovering the essence of a phenomenon. By modulating some aspects of the entity while holding others constant, we are able to discern what is essential to the entity from what is contingent about it. For example, if we take a cube, its colour is a contingent fact about it. But its having six sides is essential. By modulating each feature we can reveal what is essential about a cube and what is merely contingent. Some of these phenomenological techniques are, of course, helpful for a phenomenological analysis of illness, as will be discussed in the coming chapters.[9]

Once this suspension of ordinary beliefs, which Husserl terms the 'natural attitude', takes place, a 'presupppositionless' study of consciousness becomes possible. The suspension takes place via a process Husserl called the 'phenomenological reduction', in which reality comes to be treated as a correlate of consciousness, rather than a foundation out of which consciousness arises. Husserl viewed consciousness as the basis of all experience. But what is the relationship between consciousness (or 'soul' as he called it) and the body?

Husserl's groundbreaking analysis in Division II of *Ideas II* develops a phenomenology of the lived body as an organ of perception and as an

---

[9] See also Carel (2010) and Chapter 8, this volume.

organ of the will. Husserl bases his analysis on an understanding of the conditional dependency of bodily consciousness on material and worldly circumstances; this 'conditionality' provides a basis through which to understand the relationship between one's body and the world, and the ways in which the body not only acts on but is affected by and constrained by the world.

Husserl reveals in his analysis the modes of being of the reality of one's own body and of its manifestation. And he dedicates a large portion of the text to the question of empathy, how we perceive the expressive body of the other, and how my body appears for the other. As Rudolph Bernet writes, in these sections of *Ideas II*, Husserl offers a subtle understanding of 'the intimacy of bodily sensations and the transcendence of its insertion in the world' (Bernet 2013, p. 46).

Husserl says of the mind/body relationship: 'Body and soul form a genuine experiential unity' and that in virtue of this unity 'the psychic obtains its position in space and time' (Husserl 1989, p. 176). But as Bernet notes, 'far from being the result of a unification of two distinct heterogeneous substances, for Husserl, the unity of body and soul (*Seele*) is a *sui generis* reality in which the two levels are not only inseparable, but, for me at least (if not for others), also indistinguishable' (Bernet 2013, p. 46).

Importantly, the unity here is not between an objective body (*Körper*) and a soul, but between a lived body (*Leib*) and a soul. With this distinction between the lived and the objective body Husserl moves further away from traditional terminology towards a phenomenological understanding of the mind/body relationship. The lived body is the body as we experience it from a first-person perspective: 'that in which are localized the sensible events by means of which the sensible determinations of objects are presented to consciousness, and that in which are localized the sensible events by means of which the experiencing subject is aware of its own activities as perceptual organ' (Drummond 2007, p. 46).

Husserl offers a detailed and sensitive analysis of the kinds of sensations the body has and how these are made up of several (sometimes overlapping) dimensions: touching and being touched, feeling something and perceiving someone else feeling something, touching vs seeing, bodily spatiality, and the conditionality of the lived body on the world in which it lives. These come together to provide an account of the primacy of the lived body and perception and of the ways in which the world becomes known to us through acts of consciousness.

Merleau-Ponty's understanding of phenomenology further develops Husserl's ideas in providing a robust account of human experience as founded on perception (2012). Perception, in turn, is itself an *embodied* activity. This is not just an empirical claim about perceptual activity but a transcendental view that posits the body as the condition of possibility for perception and action. For Merleau-Ponty the body is:

> not merely one expressive space among all others . . . Our body, rather, is the origin of all the others, it is the very movement of expression, it projects significations on the outside by giving them a place and sees to it that they begin to exist as things, beneath our hands and before our eyes (2012, p. 147).

As Gallagher and Zahavi write: ' . . . the body is considered a constitutive or transcendental principle, precisely because it is involved in the very possibility of experience' (2008, p. 135).

For Merleau-Ponty, perceptual experience is the foundation of subjectivity. Our way of being is circumscribed by the types of experiences we have and the kinds of actions we perform, which are shaped by our bodies and brains. Any attempt to understand human nature would have to begin with the body and perception as the foundations of personhood (2012, pp. 146–8). This claim is a radical one in the context of Western history of philosophy, in particular the Platonic and Cartesian traditions, in which rationalism and a conception of a disembodied mind have been central (Lloyd 1984). It is also, as we shall see, crucial to understanding illness as an essentially embodied experience.

Merleau-Ponty rejects a broadly rationalist view (which he calls 'intellectualism') by emphasizing the importance and primacy of sensual knowledge and perception (Merleau-Ponty 1964b). Because of the inseparability of embodiment, perception, action, and subjectivity changes to one's body often lead to changes in one's sense of self and in one's way of being in the world. Likewise, changes to one's personhood, as for example in mental disorders such as schizophrenia or depersonalization disorder, often lead to changes to the way in which one experiences one's body (Ratcliffe 2008a; Ratcliffe 2012a).

The fundamental role Merleau-Ponty affords perception and the body give rise to his criticism of rationalist views of knowledge as conceptual and innate. But Merleau-Ponty is also dissatisfied with the broad view he calls empiricism, although his account of it is rather narrow. On his view, empiricism is unable to account for the qualitative first-person experience that arises from sensual stimuli, because of its commitment to mechanistic causal explanations. Moreover, he thinks that because it views sense data as the basic unit of experience, empiricism fails in the attempt to describe how perceptual acts take place. Seeing perception as an aggregate of discrete units of information can never yield a meaningful, ordered human conscious experience, he argues (2012, pp. 3–12).

What Merleau-Ponty offers is a 'third way', a view of the human being as essentially embodied, a 'body-subject' that arises from acts of perception. These acts are global and meaningful; they are not discrete units of data. Perceptions are 'inhabited by sense' and are always grasped as meaningful for us (Merleau-Ponty 2012, p. 52). As another phenomenologist, Martin Heidegger, puts it: 'we hear the door shut in the house and never hear acoustical sensations or even mere sounds' (1993, p. 152). Heidegger says: 'we do not throw a "signification" over some naked thing which is present-at-hand' (1952, p. 190).

Merleau-Ponty used a gestalt view to develop his notion of the phenomenal field, the horizon through which we encounter the world (2012, p. 60). When we perceive a dot, we do not perceive it on its own; it is located in a visual field, against a background. This background and perception itself are never static or passive. In one instant we may be focusing on the dot; the following moment, our eyes may move away from the dot to another visual object, pass quickly over the entire field of vision, or shut. Each of these possibilities is a possible horizon, indicating the openness and constant change of our phenomenal field. This openness is both spatial (where do we look next?) and temporal (the dot may change its appearance when night falls) (2012, p. 68).

It is also infused with emotion, which is part of the meaning of possibilities, as we find some possibilities enticing, beckoning us (what eyes can resist the glimmer of a shiny object?) while others are unappealing. And as Husserl notes in his lecture notes from 1920–6, published as *Analyses concerning Passive and Active Synthesis*, perception is not a passive taking in of information, but an active, creative, free movement that is characterized by its spatio-temporal openness, as well as by the promise of further action that opens up new perceptual possibilities (Husserl 2001, pp. 83–91).

Contemporary philosophers such as Alva Noë (2004) continue this line of thought and see perception as a species of action, or as enactive. In addition, perception gives rise not only to further perceptual actions, but also to a range of responses that may be invited by a particular perceptual act (Husserl 2001). For example, perceiving a tree gives rise to a host of potential reactions to the perception: climbing the tree, sitting under it, or chopping it down for fuel. Perception is intimately linked to possibilities for action, as well as new perceptual possibilities that are contained in nascent form in each perceptual act. The sides of the tree I cannot see from my current position are there in the background of my perceptual act, as potential perceptual objects. As Ratcliffe notes:

the kinds of possibility that entities are perceived to offer surely reflect our various practical concerns, dispositions and capacities too . . . How something is significant to us, how it 'matters' to us, is at the same time a sense of the possibilities that it has to offer. Once this is acknowledged, the range of possibilities that we are receptive to expands considerably, given that things appear to us as significant in a range of different ways (2013, p. 77).

This practical source of meaningfulness is to be found, for example, in Heidegger's (1962) tool analysis in Division I of *Being and Time*. And as Husserl writes in a famous passage from *Ideas II*:

In ordinary life, we have nothing whatever to do with nature-Objects. What we take as things are pictures, statues, gardens, houses, tables, clothes, tools, etc. These are all value-Objects of various kinds, use-Objects, practical Objects. They are not Objects which can be found in natural science (1989, p. 29).

Husserl refers here to the deep difference between everyday engagement with objects as tools or as having value (e.g. a hammer, a painting) and the scientific view of objects as entities studied in abstraction from the life world. The priority of the everyday engagement is erased in scientific study, which often presupposes the epistemic priority of the scientific view.

Both Husserl and Merleau-Ponty see the body and perception as the seat of personhood, or subjectivity: 'Body and soul form a genuine experiential unity', writes Husserl (1989, p. 176). At root, a human being is a perceiving and experiencing organism, intimately inhabiting and immediately responding to her environment. To think of a human being is to think of a perceiving, feeling, and thinking animal rooted within a meaningful context and interacting with things and people within its surroundings. As Husserl writes in *Ideas II*, 'a human being's total consciousness is in a certain sense, by means of its hyletic substrate, bound to the body' (1989, p. 160, italics removed).

The unity of mind and body is thus paramount to both thinkers. This view of the human being as a human animal (with culture, sociality, and a meaning-endowed world) sees the body as the seat and *sine qua non* of human existence. To be is to have a body that constantly perceives the world. As such, the body is situated and intends towards objects in its environment. Human existence takes place within the horizons opened up by perception. Merleau-Ponty (2012) calls this unity the 'intentional arc':

the life of consciousness . . . is underpinned by an 'intentional arc' that projects around us our past, our future, our human milieu, our physical situation, our ideological situation, and our moral situation, or rather, that ensures that we are situated within all of these relationships. This intentional arc creates the unity of the senses, the unity of the senses with intelligence, and the unity of sensitivity, and motility (2012, p. 137, translation modified).

In a normal situation, the body-subject engages in a 'primordial dialogue' with the world. This dialogue is a pre-reflective absorbed engagement with the environment, and takes place constantly in everyday activities. For example, when we go for a walk, our legs propel the body forward, the labyrinths in our ears keep us upright and balanced, and our eyes provide visual information about the path ahead and any obstacles to be negotiated. A second dialogue takes place between different body parts and types of information, allowing the body-subject to synthesize information coming from different body parts to create a unified, fluent, meaningful experience.

All the while the walker could be avidly discussing Nietzsche, paying little attention to her body, or to walking. This lack of attention does not make the walker disembodied, of course. It simply shows that embodiment is a condition of possibility for a realm of subjectivity as we know it to exist. These 'primordial dialogues' can be characterized as silent dialogue: they are tacit and enable activity, rather than being its focus.

Whether consciously experiencing bodily sensations or being preoccupied by a completely abstract mathematical problem, both activities, and the whole spectrum in-between, are possible only in virtue of existing as embodied in a world. This holds true even if no attention is paid to the body, as is often the case when one is absorbed in a task. The body in these cases is an enabler of action and experience, a condition for human experience, rather than an intentional object. Perhaps it can be said that the body is always an intentional subject, to coin a term, but only sometimes the intentional object of consciousness.

But as we shall see, even when we engage in the most abstract of activities, the body is never completely effaced; it is always felt and it can move back into being an object of attention with a sensation as slight as an itch. More substantial and prolonged types of bodily demands call attention to the body in more elaborate ways, as will be discussed in the next chapter. Such attention brings to light the body's practical and theoretical significance: the body is inseparable from, and the condition for, any experience whatsoever. Its foundational status with respect to experience must be philosophically elaborated prior to any analysis of illness.

1.4 The Habitual Body, Motor Intentionality, and Intentional Arc

We now turn to look more specifically at Merleau-Ponty's analysis of the body, and at key terms in the *Phenomenology of Perception*, which will enable us to provide an analysis of illness in the next chapter. Many of our actions, particularly everyday routine actions, are pre-reflective: they are the product of habit rather than conscious reflection. A complex web of such habits makes up our world. Our habits and ordinary ways of engaging with our environment constitute a meaningful world with which we seamlessly interact. Against this often invisible background activity, reflection and conscious thought take place. Normally we pay attention to what is preoccupying us at a given moment rather than to the cup of tea we are preparing. But Merleau-Ponty, building on Husserl, wants to direct our attention to the significance of this silent background: it plays an essential role in the constitution of subjectivity, as the previous sections discuss.

The body, for Merleau-Ponty, has a dual role. On the one hand, it is a physical thing, an object that can be weighed, measured, and described from a third-person perspective using purely physical or naturalistic terms. But the body is also the source of subjective feelings, perceptions, and sensations, the seat of subjectivity, the place where consciousness occurs. As such the body is a subject-object, a unique being that can be experienced both from a first- and a third-person point of view. As Husserl (1989) writes: 'the Body is originally constituted in a double way: first, it is a physical thing, *matter* ... Secondly, I find on it, and

I *sense* "on" it and "in" it: warmth on the back of the hand, coldness in the feet, sensations of touch in the fingertips' (1989, p. 153). The body participates in two orders: that of causation (matter) and of motivation (sense).

Merleau-Ponty further develops the example, taken from Husserl (1988), of two hands touching each other. Each hand is touching, active, and sensing the other hand but also being touched, passive, and sensed by the other hand. This simple action demonstrates the duality of the body. As Merleau-Ponty (2012) describes it:

> When I touch my right hand with my left hand together, the object 'right hand' also has this strange property, itself, of sensing … the two hands are never simultaneously both touched and touching. So when I press my two hands together, it is not a question of two sensations that I could feel together, as when we perceive two objects juxtaposed, but rather of an ambiguous organization where the two hands can alternate between the functions of 'touching' and 'touched' (2012, p. 95).

This view of the body as both an active touching subject and a passive touched object posits it as unique in nature. The body is recognized as having the power to give rise to these 'double sensations': the sensibility of the active touch and of the passive being touched. An active touch yields sensations that relate to the qualities of what I touch (hardness, warmth) and sensations that relate to the hand (or another part of my body) that touches. A passive 'being touched', on the other hand, gives rise to a simple sensation that a surface of my body feels when it is touched (Bernet 2013).

But when the body touches itself (Husserl gives the example of touching one's forehead, or two hands touching each other), things become much more complex and this dual status of the body comes to the fore. As Bernet notes, in the intersection of different sensations belonging to my two hands some of the *continuity* and coherence of my flesh is constituted. But this intersection also allows us to experience the *difference* between one's organs or parts (Bernet 2013, p. 50). As he puts it, 'my body simultaneously explores itself from the outside and feels itself from the inside' (p. 50).

The philosophical significance of these double sensations is described by Bernet as expressing the intimacy of bodily sensations and the body's insertion in the world (2013, p. 64). Although he uses the traditional terms 'body' (*Leib*) and 'soul' (*Seele*), for Husserl the unity of human beings is not the combination of a material body (*Körper*) with pure spirit (*Geist*)—body and soul are indistinguishable. The body is both an organ of perception and a lived physical body. This duality will be a prominent principle in the phenomenological analysis of illness, as we shall see in the next chapter.

Another central concept that will be of use in the analysis of illness is Merleau-Ponty's notion of motor intentionality. He challenges the view that only mental phenomena can have intentionality by extending the concept to include bodily intentionality. This is the body intending towards objects, directing itself at goals, and acting in a way that is 'about' various aims and objects. For example, if I reach with my hand to grasp a cup of tea, my hand intends towards the intentional object, the cup. The position of the hand, the direction of the movement, the curling of the fingers are all directed at, or intended towards, that cup. Motor intentionality connects my body to the cup of tea. This notion captures the intelligibility and goal-directedness of bodily movement. Through bodily intentionality we are able to make sense of a collection of disparate bodily movements, unifying them into a meaningful action (Merleau-Ponty 2012, p. 139).

In this sense, we could say that motor intentionality is an analogue of mental intentionality. But Merleau-Ponty is making the stronger claim that bodily intentionality is primary to, and the foundation of, mental intentionality. He sees motility as basic intentionality, reiterating Husserl's claim that 'consciousness is not an "I think that", but rather an "I can"' (Merleau-Ponty 2012, p. 139 and see Husserl 1988, p. 97). There can be no mental intentionality without bodily orientation in a world. 'Consciousness is being towards the thing through the intermediary of the body . . . to move one's body is to aim at things through it' (Merleau-Ponty 2012, p. 140, translation modified).

Motor intentionality is embedded within the intentional arc mentioned earlier in the chapter. The intentional arc is the term Merleau-Ponty coins to describe our relationship to the world. This relationship includes a layer of motor intentionality but also a temporal structure (compare Heidegger 1962), a human setting, and a moral and existential situation. These capture the unique relationship a human being has to the world: a relationship that is not only physical, but also embedded in cultural and social meaning and is ultimately an *existential situation* rather than a mere physical position. The intentional arc brings about the unity of the senses, cognition, sensibility, and motility—in short, the unity of experience. And as Merleau-Ponty puts it, it is the intentional arc itself that 'goes limp' in illness (2012, p. 137).

This view of human embodiment put forward by Husserl and developed by Merleau-Ponty sees the body as an intelligent, planning, and goal-oriented entity. The body is not a passive material structure waiting for mental commands, but is actively engaged in meaningful intelligent interaction with the environment. Through its directedness the body executes actions that are not merely physical movements but goal-directed movements that can only be understood as such. 'For us the body is much more than an instrument or a means; it is our expression in the world, the visible form of our intentions' (Merleau-Ponty 1964, p. 5). Thus the body is the core of our existence and the basis for any inter-action with the world; it is our general medium for having a world (Merleau-Ponty 2012, p. 147).

This view of embodiment as the fundamental characteristic of human existence is in line with recent literature on embodied cognition (Clark 1997, 2008; Wheeler 2005), enactment, and some of the attempts to reconcile phenomenology and naturalism (Petitot et al. 1999; Carel and Meacham 2013). This view has been adopted by researchers in diverse fields, including education, linguistics, ecological psychology, and artificial intelligence (Calvo and Gomila 2008; Lakoff and Johnson 1999).

Although working in diverse fields, these researchers share the understanding that 'cognition and behaviour cannot be accounted for without taking into account the perceptual and motor apparatus that facilitates the agent's dealing with the external world . . . ' (Calvo and Gomila 2008, p. 7). This broad research programme adopts (implicitly or explicitly) phenomenological ideas and methods. It regards phenomenology as a useful framework through which to think about the mind/body relationship, perception and action, embodiment, consciousness, and a wide range of issues in the philosophy of mind (Gallagher 2005; Noë 2004). The use of phenomenology to describe the experience of illness, and the emphasis on embodiment as a core feature of the human being, is part of this research programme.

Although this section distinguished phenomenology from empirical psychology, that is not to say that there are no important and bilateral interactions between the two. Many contemporary phenomenologists consider the findings of empirical science to have major importance for phenomenology and explore the relationship between these empirical domains and phenomenology as part of their philosophical endeavours (Zahavi 2013; Wheeler 2013; Dorfman 2013; Gallagher 2005; Gallagher and Zahavi 2008). Philosophers such as Michael Wheeler advocate letting phenomenology and cognitive science inform and constrain one another. He writes:

> The point at which the transcendental dimension of phenomenology meets the naturalistic dimension of cognitive science is not necessarily the site of a barrier to an alliance between these two modes of inquiry... This dynamic of selective mutual constraint and influence which characterizes this interplay means that the friction in force here is of the positive (productive) and not the negative (antagonistic) kind (Wheeler 2013).

1.5 Survey of Literature on Phenomenology of Illness

As we have seen, phenomenology engages in the study of consciousness in a very different way to empirical psychology or other sciences. And we can see how phenomenology may illuminate particular aspects of human experience, such as perception or memory. But can phenomenology be of use in the study of a concrete, socially and culturally situated, diverse and varied experience such as the experience of illness? Can illness be abstracted to core features that are experienced by all and only cases of ill health?

S. Kay Toombs, a philosopher suffering from multiple sclerosis, who has written beautifully and extensively on the phenomenology of illness, suggests precisely that. In her early work she searches for the eidetic (essential) features of illness, which characterize all illness, beyond the particular features of specific diseases (1987, p. 229). Toombs's aim is to find these eidetic features of illness, which transcend peculiar and particular features of different disease states and constitute the meaning of illness as lived.

On her view, these unvarying characteristics of particular disease states are analogous to eidetic characteristics of illness and uncovering these enables a shared world of meaning, so those who are not ill can better understand this experience. The practical usefulness of this approach is clear: it can improve patient–clinician communication, increase patient compliance and trust, assist in medical teaching and training, and enable patients to better understand and order their own experiences (for an overview of these practical applications see Carel 2010).

So what are the essential features of illness? Toombs describes five such features: loss of wholeness, loss of certainty, loss of control, loss of freedom to act, and loss of the familiar world (1987, p. 229ff.). These are further broken down into features such as bodily impairment; profound sense of loss of total bodily integrity; the body can no longer be taken for granted or ignored; the body thwarts plans, impedes choices, renders actions impossible; disruption of fundamental unity of body and self. Further features include experiencing the body as other than self; loss of faith in body; perceiving the body as threat to self; radical loss of certainty; experiencing the illness as capricious interruption; loss of control; unpredictability. And finally, the patient is isolated from the familiar world and unable to carry on normal activities; the future is truncated. For Toombs, regardless of what particular disease one suffers from, these features will be present and thus serve to reveal the experience of illness beyond its surface features which vary from one case to another. As she writes: 'the eidetic characteristics of illness transcend the peculiarities and particularities of different disease states and constitute the meaning of illness-as-lived. They represent the experience of illness in its qualitative immediacy' (1987, p. 229). This account has been very influential and used a starting point to many later works on the subject.

Other authors focus on specific features of illness that can be explored phenomenologically. They have drawn on a host of phenomenological writings, from Husserl to Merleau-Ponty, Sartre, Gurwitsch, and Heidegger. One of the earliest authors is Richard Zaner (1981, 2005), whose seminal work, *The Context of Self*, has had considerable influence on phenomenological work within nursing and healthcare research. Zaner's early work examines embodiment while taking pathology and illness as limit cases that illuminate normal embodiment (1981). His later work on narratives of illness is more implicitly phenomenological, but its starting point—seeing patients' ethical dilemmas as grounded in concrete existential situations—remains phenomenological (Wiggins and Sadler 2005). Drew Leder's (1990) notion of the absent body has been similarly influential (discussed in Chapter 2). Other notable authors are Toombs (1987, 1988, 2001) who was already discussed and will be returned to in Chapter 2, Fredrik Svenaeus (2000a, 2000b, 2001), and

Anna Luise Kirkengen (2007), each of whom developed a unique phenomenological account of illness. Matthew Ratcliffe (2008a) developed a phenomenological account of mental disorder, and is the foremost contemporary philosopher of psychiatry using phenomenology.

Fredrik Svenaeus offers a unique view of medicine's aim, using a hermeneutic phenomenological approach to describe medicine as an interpretive practice (2001). This emphasis on hermeneutic aspects of the patient–clinician encounter, as well as on the interpretative work involved in diagnosis and in other epistemic aspects of medical work, draws on Gadamer's framework to provide a view of illness as based in social and interpretative practices of generating meaning. Svenaeus views illness as an experience of uncanniness, or 'unhomelike being in the world' (2000a). Such a characterization of illness sees it as an experience of 'a constant sense of obtrusive unhomelikeness in one's being-in-the-world', in which our transcendence into the world becomes incoherent and loses its sense of order and meaningfulness (2000a, pp. 10–11). This arises from the changed conditions of embodiment, and the loss of attunement of one's body with the environment.

On this view, the role of medicine is to overcome the unhomelikeness of illness, or to help the ill person find their way home, back to a homelike being-in-the-world (Svenaeus 2000a, pp. 10–11). Thus medicine becomes 'the art of providing a way home for the patient' (p. 14). The phenomenological approach supports Susan Sontag's (1978) claim that illness is not merely a metaphor, narrative, or story. Illness is primarily a bodily experience that gains meaning from social and cultural context, but is first and foremost lived as a bodily experience of suffering and limitation.

Phenomenological concepts such as being-in-the-world, authenticity, anxiety, uncanniness, and the body-as-lived have been used to describe the experience of illness (Svenaeus 2000b; Toombs 1987, 1993, 2001; Carel 2013a). Viewing illness as transforming one's being-in-the-world, including one's relationship to the environment, social and temporal structures, and one's identity, has helped capture the pervasive nature of illness. Some authors have tried to identify health with authenticity and illness with inauthenticity, although this approach has been widely criticized (Svenaeus 2000b; Keane 2014; Withers, unpublished dissertation).

Philosophical work on anxiety has been extremely productive in developing an account of 'existential feelings' and of how anxiety (and other mental disorders) involves pervasive symptoms that cut across the psyche/soma distinction (Ratcliffe 2008a; Carel 2013b). The distinction between the biological body ('objective body') and the body-as-lived ('subjective body') has been utilized to express the difference between disease and illness, as well as to account for the difference in perspective between clinician and patient (Toombs 1987; Carel 2013a).

Work has also been done to bring together biological and phenomenological understanding of trauma in order to develop a holistic account of the ways in which such events affect human beings (Getz et al. 2011; Kirkengen 2007). These authors have developed a rich and valuable phenomenological treatment of illness. However, further work is needed in order to understand specific aspects of the experience of illness, such as bodily doubt (see Chapter 4, this volume) and to describe particular conditions and situations of illness.

In addition, no recent account has attempted to reconcile the conflicting findings of existing qualitative studies of the experience of illness. Some studies provide evidence that illness impacts negatively on well-being, while others show that it does not (although Thorne et al. published a meta-study in 2002). The plurality of concepts, traditions, approaches, and accounts all belonging to the phenomenological branch of qualitative research have been criticized as too promiscuous to constitute a unified qualitative research methodology (Earle 2010). These challenges are fundamental to this book's goal, which is to provide a comprehensive and coherent phenomenology of illness. The question whether well-being within illness is possible is taken up in Chapter 6.

Another type of challenge has been posed by authors who criticize the view that illness experiences have general and universal essential features, as Toombs (1987) suggests in her early work. On their view, the specificity of particular conditions and the ways in which they are debilitating, as well as the concrete context in which illness is experienced, cannot be stripped away from the experience of illness. This view is critical of Husserl's construal of phenomenology as a transcendental science of consciousness, claiming that the abstraction of concrete contents of an individual life removes much of what is essential to it (Merleau-Ponty 2012).

This view suggests using phenomenology to study the specific details of particular illness situations (e.g. 'the experience of stroke in middle-aged men'). This has been developed into a qualitative research method, but has also come under criticism, for losing sight of phenomenological principles and becoming indistinctive from other methods of qualitative research which also rely on first-person reports (Earle 2010). Other aspects of the experience of illness that have been studied phenomenologically have been the patient–clinician relationship (Carel and Macnaughton 2012; Toombs 1989, 2001), temporal experience in illness (Toombs 1990), the phenomenology of mental disorder (Ratcliffe 2008a; Stanghellini 2004), and the role of shame in the medical encounter (Dolezal 2015).

1.6 Conclusion

This chapter presented phenomenology, with particular attention to the mind/body relationship, intentional arc, motor intentionality, and the habitual body. It anchored an understanding of the human being as embodied in this phenomenological framework, and explained what embodiment means in a phenomenological context. It also discussed the disease/illness distinction, and asked how the distinction applies to mental disorder. Finally, the chapter provided an overview of phenomenological literature focusing on illness. With this general framework in place and the literature survey providing some orientation in this growing field, we now turn to developing a phenomenology of illness, the task of the following chapters.

2

Phenomenological Features of the Body

So far I have suggested that phenomenology can be used to describe illness by focusing on first-person accounts of what it is like to suffer from a particular condition.[1] On Merleau-Ponty's view, our experience is first and foremost an embodied experience, an experience of fleshly sensual existence. So a deep and permanent change to the body, such as takes place in serious chronic illness, would lead to far-reaching changes to one's embodied experience. Thus phenomenology seems doubly suited for describing the experience of illness, which often includes a radical shift in one's embodiment.[2] It provides a framework that gives detailed attention to experience. And it takes as its starting point the centrality of embodiment and of perception.

However, such an analysis is a challenging undertaking. The experience of illness is diverse and constantly changing; it is bound up with cultural and personal meaning; it can be radically subjective and difficult to describe, or possibly unshareable in some respects, as Toombs and others claim (Toombs 1993, p. 23; Carel and Kidd 2014). And yet, such an analysis seems essential to our quest to understand illness. When we think about a phenomenological description of illness, immediate questions arise: do illness experiences share certain general features? Are these features universal or essential, or are they culturally variable? Do different illness experiences, such as the experience of acute vs chronic

---

[1] It can be used to understand any type of bodily experience, e.g. Young's phenomenological analysis of the embodied experience of pregnancy (Young 2005b).
[2] First-person experience is also useful for a number of other issues in medicine, for example medical training, simulation of particular illnesses (e.g. simulating angina using a chest band), and understanding the experience of providing care in a health setting, to name a few.

illness, share some features? Do mental disorder and somatic disease have any common features?

Addressing these questions is the task of this chapter and the next. This chapter offers a phenomenological framework through which to study illness, while the next chapter puts forward a detailed analysis of the dimensions of illness. The framework developed in this chapter consists of five interconnected themes that offer a nuanced understanding of the experience of illness. The first is Toombs's analysis of the features of illness. The second is the phenomenological distinction between the objective body and the body as lived. The third is Sartre's three orders of the body. The fourth is the claim that the healthy body is transparent and illness is the loss of this transparency. Finally, using Heidegger's tool analysis, I suggest a fifth theme: illness is a breakdown of 'bodily tools'.

2.1 The Features of Illness

Earlier, I described the challenge of understanding illness as stemming in part from the diversity and variation between and within different people's illness experience. There are tens of thousands of diseases, and even if we take the same illness, each person is affected differently by symptoms, prognosis, pain, and so on, both emotionally and physically. It may therefore seem like an impossible task to try to distil the shared features of illness that characterize the illness experience generally. However, phenomenologist S. Kay Toombs (1987), herself a sufferer of multiple sclerosis, has performed such an analysis. In adopting a phenomenological approach, Toombs claims that although the experience of illness is complex, it nonetheless exhibits a typical way of being (Toombs 1987, p. 228; Toombs 1992, pp. 90–8).

Certain features of illness are manifest regardless of the particular disease state. These, claims Toombs, are the invariant characteristics of illness; in her later work Toombs (1993) calls them typical characteristics, rather than invariant, to indicate her move away from a Husserlian framework. These characteristics are integral to the illness experience and remain at its core, regardless of varying empirical features (Toombs 1987).

Toombs lists five typical characteristics of illness: the perception of loss of wholeness, loss of certainty, loss of control, loss of freedom to act, and loss of the familiar world. These losses represent the lived experience of illness in its qualitative immediacy and are ones that any patient, in whatever disease state, will experience. Cumulatively, they represent the impact of the illness on the patient's being-in-the-world.

Toombs begins with the loss of wholeness. This loss arises from the perception of bodily impairment, which leads to a profound sense of loss of bodily integrity. The body can no longer be taken for granted or be seen as transparent or absent (compare Leder 1990), as it assumes an opposing will of its own, beyond the control of the self. The ill body thwarts plans, impedes choices, and renders actions impossible. In addition, illness disrupts the fundamental body–self unity, and the body is now experienced as other-than-me (Toombs 1987). Thus illness is experienced as a threat to the self, so the loss of integrity is not only of bodily integrity, but also of the integrity of the self (Toombs 1987, p. 230; Toombs 1992, pp. 90–2).

The second kind of loss, the loss of certainty, ensues from the loss of wholeness. The patient 'is forced to surrender his most cherished assumption, that of his personal indestructibility' (Toombs 1987, pp. 230–1; Toombs 1992, pp. 92–4). This forces the individual to face her own vulnerability. The recognition of vulnerability and loss of certainty causes great anxiety and worry and this deep apprehension is difficult to communicate. Illness is experienced as a 'capricious interruption': an unexpected mishap in an otherwise carefully crafted life.

This experience of illness as an unexpected calamity leads to a sense of loss of control, the third kind of loss Toombs describes. The illness in its seemingly random unfolding (will the cancer cells respond to the chemotherapy? Why did I suffer the heart attack?) is experienced more like a stroke of bad luck than freely chosen life circumstances. This makes the familiar world suddenly seem inherently unpredictable and uncontrollable (Toombs 1987, p. 231; Toombs 1992, pp. 93–4). This leads to a further heightening of the sense of loss of control caused by the realization that the belief that medical science and technology protect us from the vagaries of ill health is nothing more than an illusion harboured by modern man. In addition, the ill person's ability to make rational choices is eroded because of her lack of medical knowledge and limited ability to judge whether the health professional professing to heal can in fact do so (Toombs 1987, p. 232; Toombs 1992, pp. 95–6).

This leads to the fourth kind of loss: the loss of freedom to act. The ill person's ability to choose freely which course of action (which medical treatment) to pursue is restricted by her lack of knowledge of what the best course of action may be.[3] Moreover, in deciding whether to accept medical advice, the patient often assumes that the physician understands and shares her value system and takes these values into account when recommending a certain course of action. However, the physician may often feel that it is inappropriate, irrelevant, or intrusive to inquire about the patient's values, and judges the clinical data alone to be sufficient for determining what is best for the patient. 'Thus patients not only lose the freedom to make a rational choice regarding their personal situation but additionally lose or abrogate the freedom to make the choice in light of a uniquely personal system of values' (Toombs 1992, p. 96).

Finally, the fifth kind of loss, the loss of the everyday world, arises from the disharmony illness causes and its distinct mode of being in the world (Toombs 1992, p. 96.). The ill person can no longer continue with normal activities, or participate in the world of work and play as before. Whereas friends and colleagues live as they did previously, the ill person's familiar world is often lost to the demands of disease, and this difference exacerbates the sense of loss. A large part of the familiarity of the world arises from its sharedness with other people, which is also lost in illness, when people's ability to do things is affected and they can no longer play sports or stay up late, for example (Toombs 1992, p. 97). The temporal dimension of one's world is also shaken because future plans have to be adjusted in light of a medical prognosis and the healthy past is broken off from the ill present (compare Bury 1982). 'The future is suddenly disabled, rendered impotent and inaccessible' (Toombs 1987, p. 234). And this loss of future further isolates the ill person from her hitherto familiar world.

Once shattered, all of these domains (wholeness, certainty, control, freedom, and familiarity) can only be re-established tenuously. But even if the losses are restored, any such re-establishment is always

---

[3] Many of our decisions are made under epistemic uncertainty and varying degrees of ignorance. This has been recently explored by L. A. Paul (2014) who argues that certain experiences are transformative in nature, thus posing a challenge to the view that decisions can be made rationally, i.e. taking into account one's future preferences and beliefs. I suggest that illness experiences may be transformative in Paul's sense, as I discuss in Chapter 6, this volume.

accompanied by a sense of fragility and uncertainty, a sense of tenuousness (Carel 2013b). In many ways the process Toombs describes is irreversible, even if health is largely restored. That is an additional reason why Toombs focuses on these features as fundamental to the experience of illness and claims that these features pertain regardless of the particular disease state of the individual. These characteristics represent 'the "reality" of illness-as-lived. They reveal what illness means to the patient' (Toombs 1987, p. 234). For Toombs, this model of illness makes the primacy of the person explicit, and not secondary to an objective, abstract, disease entity, as the biomedical model has it. In this sense the phenomenological model of illness can better serve not only patients, as it would naturally seem, but also physicians, whose ultimate goal is to improve individual patients' lives, not merely treat a disease process.

Let me make a few comments on Toombs's account. First, it seems that the loss of freedom is broader than the loss Toombs describes in her early work. In that work, Toombs focuses on the loss of freedom to make rational decisions on the best course of action in response to the medical facts. However, this loss is only one in a much broader loss of freedom brought about by illness. The loss of bodily freedom, freedom to make life plans, and freedom from anxiety about one's body is acute in both somatic and mental illness. Arthur Frank, for example, describes his diagnosis of cancer in these words: 'What was it like to be told I had cancer? The future disappeared. Loved ones became faces I would never see again. I felt I was walking through a nightmare that was unreal but utterly real. [...] *My body has become a kind of quicksand*, and I was sinking into myself, into my disease' (1991, p. 27, emphasis added).

A similar closure of the future and of the freedom to choose (to an extent) one's course of action and future goals, is also a prominent theme in mental illness. John Stuart Mill, who suffered from depressive symptoms, describes his illness in his autobiography: 'the whole foundation on which my life was constructed fell down [...] The end has ceased to charm, and how could there ever again be any interest in the means? I seemed to have nothing left to live for' (1989, p. 112). In a situation of acute dejection the freedom to pursue goals is effaced by the loss of meaning of any goal. Although he is free, Mill cannot seize any particular goal because of his underlying feeling that the realization of any goal would be pointless and would not bring him happiness.

So it seems that the loss of freedom is a pervasive loss, spanning the freedom to choose one's future, but also a loss of freedom in the present, in that many routine activities easily performed are no longer possible and must be either given up or replaced by an alternative habit or a different way to perform an old task (Carel 2009).

Toombs has herself broadened the conception of the loss of freedom in her later work (1993), where she describes how in illness bodily intentionality is frustrated, the relation between the lived body and the environment is changed, and how possibilities for action shrink. The more developed account of loss of freedom in her later work stems from a more pronounced focus on embodiment as the source of meaning and locus of selfhood. On this view, illness disrupts the fundamental features of embodiment (which Toombs takes to be: being in the world, bodily intentionality, primary meaning, contextural organization, body image, and gestural display) and consequently, illness is experienced as a chaotic disturbance and sense of disorder (Toombs 1993, p. 70). As Toombs writes, 'the possibilities for action shrink. If I am ill, I simply do not have available to me all the alternatives that are available in health' (Toombs 1993, p. 63).

Second, not all losses are experienced by the ill person. For example, someone with a profound learning disability may not know that they are incapacitated by it and may not experience the losses Toombs describes (even if, in fact, the losses are real). Similarly, someone in a persistent vegetative state may experience none of the losses, despite the fact that she has, in fact, incurred them. Such examples demonstrate that the features of illness Toombs describes as characteristic of the illness experience are contingent upon the ill person's situation and capacities. One might also say that in that case such people are not ill, but merely diseased. If there is no experience of the deficit, there can be no experience of loss and hence no illness experience.

Third, an analogous argument can be made about cultural difference. It may be that in some cultures certain losses are not experienced because some values (e.g. freedom) do not exist in those cultures, or do not carry the same significance as they do in other cultures. So these features of illness should be understood in a more restrictive sense as not entirely eidetic, but as offering a general characterization of the experience of illness as lived by conscious adults with a certain degree of self-awareness, in Western societies.

In what follows I use Toombs's helpful analysis in this more restrictive sense. This analysis will provide a framework for understanding the changes illness brings about in the lives of the ill person and those around her; we will return to this framework in the following chapters. We also need a rich account of the body as it is experienced by the ill person and by others. We now turn to Sartre for that.

2.2 The Objective Body and the Body as Lived

A useful tool for a phenomenology of illness is the distinction between the objective body (which Husserl called *Körper* and Merleau-Ponty called *le corps objectif*) and the body as lived (*Leib* and *corps proper*, respectively). The objective body is the physical body, the object of medicine: it is what becomes diseased. Sartre calls this body the 'body of Others' (*le corps d'autrui*): it is the body as viewed by others, not as experienced by me (Sartre 2003). The body as lived is the first-person experience of this objective body, the body as experienced by the person whose body it is. And it is on this level that illness, as opposed to disease, appears.[4]

This distinction is fundamental to any attempt to understand the phenomenon at hand: the ill person is the only one who experiences the illness from within, although others may have an experience of someone else's illness. Only the ill person can definitively say if they feel pain or what a medical procedure or a particular symptom *feels like*. Of course we are able to surmise much from other people's behaviour; sometimes we can know better than them. For example, we can tell from a toddler's grumpy behaviour that she is tired, whereas she may not be able to formulate this for herself. However, in cases of conscious adults, we usually take each person to be the ultimate authority on his or her own sensations, feelings, and experiences. Because of this first-person authority the experience of illness contains a measure of incommunicability that should be acknowledged (Carel 2013a; Carel and Kidd 2014). Or, as Sartre put it more strongly, 'the existed body is ineffable' (Sartre 2003, p. 377).

Disease, on the other hand, is a process in the objective body that may be observed by any other person and may yield information that is not

---

[4] Some authors (e.g. Twaddle 1968; Hofmann 2002) suggest a third category, sickness, to denote the social dimension of human ailment.

available through first-person reports. For example, one may have elevated cholesterol or blood pressure, or an early stage of heart or renal disease, while having no experience of these. Often such knowledge comes from medical tests that yield objective facts with no experiential correlate. For example, elevated blood pressure may not *feel like* anything. And it is only once it is revealed via a blood pressure test that it begins to feature in the diseased person's experience.

However, the contrast between the first- and third-person perspectives is more nuanced and complex in two ways. First, there is also the second-person perspective on illness. And second, the third-person perspective is not necessarily an impersonal one. There is a second-person phenomenology of illness, which involves perceiving aspects of what the ill person experiences. The claim that we can experience something of others' experience, but in a second-person rather than the first-person way, is made by Edith Stein. For Stein, empathy

not only allows me to see or understand the inner life of another and the body of the other as both other and similar to my body, but it also permits me to see what I am not insofar as the other is not simply a modification of my own ego, my own person (Calcagno 2007, p. 65).

For example, in a case where one's spouse becomes chronically ill and increasingly limited in what she can do, it is 'our life', 'our projects', 'our meaningful activities' that are curtailed, rather than just those of one partner. There is a sense in which both partners have an illness experience, albeit an importantly different kind of experience. I further suggest that the health professional can adopt a second-order perspective towards illness, and that this perspective can be cultivated through phenomenological practices such as the 'phenomenological toolkit' discussed in Chapter 8.

The relationship between illness and disease is not simple: the two aspects do not simply mirror one another. Illness may precede one's knowledge of one's disease: disease is commonly diagnosed following the appearance of symptoms experienced by the patient. These symptoms are part of her illness experience. Disease may appear without illness (as in the example of high blood pressure with no symptoms). Or often we have both illness and disease, but the two do not perfectly cohere.

For example, severe disease or disability (e.g. quadriplegia, advanced chronic obstructive pulmonary disease (COPD)) may give rise to an illness experience that is tolerable or even experienced by the ill person as not causing severe incapacitation or suffering, due to adaptation (Carel 2009; Haidt 2006). So although the disease may be classed as severe on some clinical scale, the illness experience is not as correspondingly negative as might be expected. This issue will be discussed in detail in Chapter 6, so I will only make a few brief comments about it here.

The gap between objective and subjective assessments of health and of illness are of clinical importance because interventions ought to address patients' lived experienced but are often designed to restore objectively measured function. For example, a study may state as its end points an increase in a lung function value, or a reduction in hospital admissions. But these do not necessarily correlate with a perceived increase in health or well-being by the patient. Interventions often aim at disease without giving due attention to the fact that the relationship between disease and illness is complex, non-linear, and poorly understood.

The relationship between disease markers (disease stage, objective dysfunction, etc.) and the experience of symptoms and illness is not clearly correlated. A surprising lack of correlation between disease severity and patient well-being is well-documented (Angner et al. 2009; Carel 2007, 2009; Gilbert 2006). An example of this is a study comparing renal patients undergoing haemodialysis with healthy controls (Riis et al. 2005). We would expect the renal patients who are tethered to a dialysis machine three times a week, unable to travel and often incapacitated, to be markedly less happy then the healthy controls. In fact, both groups report a similar level of well-being (2005, p. 6).[5]

Despite the reported similarity in the two groups' levels of well-being, both the dialysis patients and the healthy controls overestimate the impact of haemodialysis on well-being. Both groups thought that the dialysis affected patients' well-being more strongly than it actually did. Other examples of this phenomenon, termed the 'disability paradox', abound (see for example Albrecht and Devlieger 1999; Ubel et al. 2005; Gilbert 2006). One way of explaining these findings is by appealing to a lack of correspondence between disease and illness, which itself arises from the underlying difference between the body as object and the body as subject.[6]

---

[5] See Chapter 6 for detailed discussion of this issue.

[6] Merleau-Ponty (2012) helpfully terms the latter the 'body-subject', to indicate the inseparability of the body from the subject under this mode.

Another reason that the difference between the objective body and the lived body emerges in illness is that the latter is in large part habitual. It is used to performing certain tasks with ease, as discussed in Chapter 1. Routine actions can be performed expertly and efficiently because they have become habit, and form what Merleau-Ponty (2012) calls the 'habitual body'. While getting ready to go to work, one rarely notices the multitude of actions and the expertise required to have a shower and get dressed. Only when we watch a novice, e.g. a child learning to button her shirt or tie her shoelaces, can we appreciate the complexity of the activity and the expertise it requires. The ease with which we perform habitual tasks often disappears in illness, where novice-like behaviours appear as a result of lost capacities (e.g. a stroke victim relearning to talk or read) and the need to find new ways to perform routine tasks. While retaining the know-how, the ability to carry out an action is lost. The habitual body loses its expert performance skills and these have to be replaced or modified. Illness thus also reveals the difference between the objective body and the habitual body.

Another example given by Merleau-Ponty (2012) to illustrate (among other things) the habitual body is the phantom limb. A phantom limb is the sensation emanating from a limb that has been amputated. The phantom limb feels painful or itchy, but the real limb has been removed. Merleau-Ponty explains the phantom limb as a rift between the objective body and the lived experience of it. The objective body has no limb, but the lived body experiences that limb as present because the body schema which contained four limbs still structures the amputee's sensations. The phantom limb is the expression, based on years of having a body schema and associated habits with four limbs, of the body as it used to be. The habitual body thus becomes a relationship to an environment and to a set of abilities that are no longer available to the amputee. 'To have a phantom limb is to remain open to all of the actions of which the arm alone is capable and to stay within the practical field that one had prior to the mutilation' (2012, p. 84).

Another example of the rift between the objective body and the lived body can be seen in anorexia nervosa. If we look at the objective body, we may see a skeletal, emaciated body. This is the objective body whose thinness can be measured by weighing it or calculating its BMI. But if we ask the anorexic to describe her body, she may say that she experiences it as obese and cumbersome. Denying this experience by making an appeal to objective facts is unhelpful. On some views of anorexia, the rift between the objective body and the body as it is experienced is the crux of the disorder (see also Bowden 2012).

The distinction between the objective and the lived body is useful in several respects: it makes clear the fundamental difference between the two perspectives. The perspective of the physician (or family members, strangers, friends) means that they can only perceive the disease through second- or third-person observation, and not from the first-person perspective, although they may, of course, care deeply about that perspective and come to inhabit a closely linked perspective.

The illness experience in its first-person form is not accessible to the physician, by definition, other than via the patient's account, when it becomes a second-person report. The patient is the only one to whom the full subjective experience of illness is available. This epistemic restriction may lead the physician to seek to treat the disease, sometimes with inadequate understanding of the illness, or to have little understanding of the impact of the illness on the patient's life as a whole (this will be discussed in the next chapter).

The patient, on the other hand, can observe the objective indicators of disease (e.g. blood test results or an X-ray) but also has unique access to the lived experience of the disease, namely illness. In this sense the patient may have, at least in principle, a unique contribution to the epistemic effort of trying to understand and treat her illness. She has access to her own illness experience *and* to the objective knowledge about the disease. But this epistemic advantage often goes unacknowledged and the patient experience may be subsumed under the medical view or discounted because the patient has no formal medical training (Carel and Kidd 2014). However, sometimes the first-person perspective may put the patient at a disadvantage. Consider the case of a patient at a terminal stage of illness, who is unable to face the fact that she is dying and thus denies it. In such a case, the health professional may have an advantage.

The claim here is that the unique ability to oscillate between the two perspectives gives the patient a deeper understanding of the illness experience, and potentially to the dual nature of the body. I do not suggest that patients have an overall epistemic advantage, as they usually lack medical training and knowledge and have no experience of the illness other than their own. But I do claim that the patient knowledge itself, which may include elements of knowing-that and of knowing-how, has not penetrated medical practice sufficiently and remains marginal to the epistemic endeavour of diagnosing and treating ill health.

The difference between illness and disease may also cause confusion and miscommunication. As Toombs (1987) notes, the physician's focus on disease may clash with the patient's primary interest in her illness, so although they may seem to speak of the same entity, they, in fact, refer to two different entities (disease, illness) and therefore have a communicative and interpretative gap that must be addressed before effective communication becomes possible.

This gap is not merely a linguistic divergence in which the two terms refer to the same object but the sense is different. Both the sense and the reference are different and thus lead to different intentionalities and modes of engagement. The health professional may focus on the disease as a physical entity while the patient's reference is her lived experience of the disease. As Toombs states:

what the phenomenological approach is concerned to show is not simply that the patient's experience should be taken into account as a subjective accounting of an abstract 'objective' reality, but that the patient's experiencing must be taken into account because such lived experience represents the reality of the patient's illness (Toombs 1987, p. 236).

2.3 Sartre's Three Orders of the Body

Seeing the health professional as occupying an objective stance while the patient lies in the subjective one helps us understand communication problems between the two parties. However, it is important to acknowledge, as was done in the previous section, that the health professional's supposed objectivity is a more complex position. Here I explore this complexity further, using Sartre's analysis of the three orders of the body.

The complexity of the health professional's stance stems from the fact that the body of the health professional and the body of the patient exist under both the objective (material, physical) and the subjective (experienced, first-person) order. Moreover, both are revealed to each other as belonging to both orders. This gives rise to a third order in which the body partakes: the order of intersubjectivity, or my body as I experience it as reflected in the experience of it by others. 'I exist for myself as a body known by the other', writes Sartre (2003, p. 375). His example is being self-conscious about one's body. This is only possible, he claims, because of this third order. The shy person is 'vividly and constantly conscious of his body not as it is for him but as it is *for the Other*' (2003, p. 376). The uneasiness the shy person feels is 'the horrified metaphysical apprehension of the existence of my body for the Others' (2003, p. 376). Only my body as it is for another person can embarrass me, not my body as I exist it. So, when a patient feels self-conscious, it is as a body experienced in this third order, the order of the body experienced as socially perceived.

Sartre discusses alienation, embarrassment, and social unease, claiming that in these situations I experience my body *as it is experienced by another*, not in the natural pre-reflective way I usually experience it. I may then begin to treat my body—or indeed to experience it—not as my pre-reflective opening to the world, but as an object that can be worked on, changed, and assessed in modes suited to objective bodies. A prominent example is reality TV 'extreme makeover' programmes, in which the attitude to the show's participants' bodies reflects this objectifying mode of self-regard.

The person seeking plastic surgery allows the surgeon to draw on her body with a pen, marking out 'excess' flesh or skin, that needs to be removed like excess clay on a sculpture. One may speculate that this may stem from the patient's experience of her body. She does not experience that flesh as part of 'my body as I exist it', but as inert matter that occludes 'my real figure'. Some plastic surgeons trade on this alienation when they accept their patients' projection of fantasies of acceptance and omniscience onto the carving of their own flesh as good reason to undertake the surgery.[7]

It is not only the patient's body, but also the health professional's body that falls under the duality captured by the disease/illness distinction, as well as the third order of the social body articulated by Sartre. The health professional may experience herself as a subject examining an object (the patient's physical body), but the object can touch back. When the

---

[7] Louis Theroux's documentary film *Under the Knife* shows patients and surgeons with such attitudes (broadcast on BBC 2, 7 October 2007 <http://www.bbc.co.uk/programmes/b008258x> accessed 19 September 2015).

physician's examining hand is 'touched back' by a body responding to its touch with a quiver or a tensing of muscles, we no longer have a subject touching an object, but a subject, which is itself also an object, touching an object, which is itself also a subject (Carel and Macnaughton 2012).

Following Sartre (2003), we can see that this seemingly simple situation contains within it a complex nexus of relations, which is the foundation of human sociality. The recognition of myself as subject for myself and as object for others is elaborated in the next step in the dance of reciprocity: the recognition of the other as object for myself and as subject for her. I meet the other *both* in her object-making subjectivity *and* as an object (Sartre 2003, p. 377). These positions are not fixed and constrained by some a priori stipulation. On the contrary: the oscillation between perceiving myself as a subject that has been objectified (the patient), which is then resubjectified in the act of touching back, continues as long as the intersubjective interaction continues. Subjectivity is forever challenged and then reclaimed, only to be challenged again.

Similarly for the physician to take an objective stance, only to have it punctured by subjective feelings, emotions, biases, that subjectify it and yet need to be checked and held back, is also an ongoing process. The objectified patient does not experience her body as an object; instead, she experiences it as 'the flight of the body which I exist' (Sartre 2003, p. 378).[8] The sense of discomfort, self-consciousness, alienation, does not arise from my being objectified *qua* the diseased body of a patient, and thus becoming an object for myself, but from the escape, or draining away, of my being *qua* subject, dissipated by the objectifying medical gaze.

The complexity arises from the body's unique metaphysical position: it is 'a non-thingly living flesh' which is neither purely an object nor pure consciousness (Moran 2010, p. 42). This intersubjective dimension of one's experience of oneself and of the other, as well as the other's experience of me and of herself, relies fundamentally on empathy. Husserl and Merleau-Ponty agree that intersubjectivity depends on empathy, which in turn depends on intercorporeality, the shared corporeal foundation underpinning our existence as individuals and as social beings (Csordas 2008). Intercorporeality itself arises from my experience of my own body as partaking in the two different orders, as

---

[8] The health professional may also be objectified, of course, but more often it is the patient whose body is scrutinized as the object of medicine.

shown in Husserl's analysis of 'double sensations' (Moran 2010, p. 41; Bernet 2013). Empathy depends on intercorporeality because fundamentally I perceive others as bodies (that are similar to mine in that they, too, sense, perceive, etc.) and I am perceived by others as a body that is similar to theirs. I am there for others and this being-there 'is precisely the body', writes Sartre (2003, p. 375).

However, the ways in which my body exceeds the first two orders (objective and subjective) and enters the third social order, are the ways in which 'my body escapes me on all sides' and returns to me as gazed upon by others (Sartre 2003, p. 375). My body is my point of view, but it is also a point of view on which other points of view are brought to bear, including points of view I could never take (p. 375). In other words, the lived body encompasses not only one's experience but also the social aspect of one's experience of one's own and others' bodies, as well as how others' experience of one's own body might impact on their own experience of their body. The experience of empathy is fundamental to this exchange and requires careful examination to see whether a radically different lived body experience may modify or curtail empathy with others or even self-empathy. We will come back to this question in the next chapter.

To end this section, it is important to return to another possible mode of interaction between health professional and patient, discussed above: that of the second-person perspective. It may be that I, as the mother of a child with chicken pox, do not have access to his itching and pain, but as someone witnessing the suffering of a loved one I do not occupy the objective observer's position. It is you, your suffering, not the objective registering of symptoms, I experience. Although I have no direct access to your pain, I am still able to empathize with it through memory, imagination, or authentic being-with. The demand to recognize the other's humanity and animality, their capacity for suffering and our shared mortality is present in the second-person stance: the I–thou relationship, or the face-to-face encounter and their ensuing ethical demands stem from the recognition of the uniqueness and irreducibility of each person (Buber 2010; Levinas 1969).

The somewhat simplistic picture of the health professional as subject who objectifies the patient is complicated by the oscillation of roles arising from the fact that we each partake in the three orders of the body. In some quarters of the medical profession objectivity is viewed as the ideal stance of the medical practitioner (personal communication during shadowing). I suggest that this view could be tempered by the possibility of the second-person perspective available to the health professional:

By recognising each other's subjectivity both physician and patient stand to gain. The physician gains a more natural mode of expression, and the patient has a feeling of being listened to by a fellow human being who neither purports to stand in her shoes, nor to be completely objective (Carel and Macnaughton 2012, p. 2335)

2.4 The Transparency of the Body

In the smooth everyday experience of a healthy body, the body as object and the body as subject are aligned and experienced as harmonious. We do not experience the difference between the two orders most of the time; they cohere and make sense as a whole.[9] The fundamental bodily experience of health is one of harmony, control, and predictability. This has led some authors to describe the healthy body as *transparent*: we do not experience it explicitly (although we do implicitly) or thematize it as an object of our attention, nor does it play centre stage in our actions, even if those actions are explicitly physical.

When writing a letter, we do not pay attention to the pen as long as it is functioning. Similarly, we do not normally pay attention to the hand gripping the pen and writing. Our attention is focused on the task we are engaged in: writing the letter. Or take a more explicitly embodied activity: if I prepare myself to catch a ball thrown towards me, I do not focus on my body but on the ball, trying to anticipate its trajectory and possible point of landing; my body simply 'follows' me to that point, arms stretching towards the ball. In such everyday experiences the physical body is not prised apart from the lived body, and the experienced functioning of the body is natural, pre-reflective, and either effortless or the effort is experienced as normal.

Sartre and Leder describe the healthy body as transparent (Sartre 2003) or even absent (Leder 1990). Sartre says: 'consciousness of the

---

[9] Although we do experience the difference between the first and third order in some social situations, when we experience our body as experienced by others. Sartre (2003) discusses shame as one such example.

body is lateral and retrospective; the body is the *neglected*, the "*passed by in silence*" ' (2003, p. 354). And Leder writes: 'while in one sense the body is the most abiding and inescapable presence in our lives, it is also essentially characterized by absence. That is, one's own body is rarely the thematic object of experience' (1990, p. 1).

The healthy body is transparent, i.e. taken for granted. This transparency is the hallmark of health and normal function. We do not stop to consider it because, as long as everything is going smoothly, the body remains in the background, the vehicle through which we experience, but not the thematic focus of experience. 'The body tries to stay out of the way so that we can get on with our task; it tends to efface itself on its way to its intentional goal' (Gallagher and Zahavi 2008, p. 143).

This does not mean that we have no experience of the body but, rather, that the sensations it constantly provides are neutral and tacit. An example is the sensation of clothes against our skin. This sensation is only noticed when we draw our attention to it or when we undress (Ratcliffe 2008b, p. 303).

There is a further sense in which we might say that the healthy body is transparent. It is not only taken for granted, but also taken to be simple, plain, or even 'empty', because if something is transparent, it is not worthy of careful observation. This view contrasts with the phenomenological emphasis on the importance of careful attention to our experience. In healthy everyday experience not only the body is overlooked, but this need itself. And of course something that seems clear or transparent at first glance may on a second, more careful, look turn out not to be clear at all. We have such opportunities for second glances not only in illness but also in a variety of everyday experiences of the healthy body.

Although we may have moments of explicit attention to the wellness of our body, for example, when a headache goes away or while exercising, it is primarily when something goes wrong with the body that it moves from the background to the foreground of our attention. When functioning normally, our attention is deflected away from our body and towards our intentional goal or action. It is not that the body is absent but, rather, that our experience of it is in the background while the object of our focus is in the foreground. 'The body is in no way apprehended for itself; it is a point of view and a point of departure' (Sartre 2003, p. 355).

In contrast, when we become ill our attention is drawn to the malfunction, which becomes the focus of attention. The harmony between the objective body and the body as lived is disrupted. Leder contrasts the healthy, absent body with illness and other situations when the body becomes an explicit object of negative attention and appears as a 'dys' (function) of sorts. 'In contrast to the "disappearances" that characterize ordinary functioning, I will term this the principle of *dys-appearance*. That is, the body *appears* as the thematic focus, but precisely as in a *dys* state [. . .]' (Leder, 1990, p. 84). The body can appear as ill, disabled, aesthetically flawed or socially awkward, objectified or sexualized, or as attracting negative attention from others, as in the experience of shame discussed by Sartre (2003).

This alleged transparency of the healthy body is somewhat idealized in philosophical descriptions of health, since it is often pierced by experiences in which the body comes to the fore, sometimes in negative ways. The first kind of experiences is social experiences of one's body as it is perceived or objectified by others. Sartre's famous analysis of the gaze (or look) as annihilating my subjectivity and objectifying my body, which becomes an object in the other's (the subject's) field of vision (2003, p. 276ff.) itself recognizes the tension between the naïve unthematized body and the social body.

Transparency is lost in any encounter in which the other's gaze posits a subjectivity within which my own subjectivity is transcended, subsumed. This is 'transcendence transcended': my own being as transcendence is transcended by another consciousness, and that realization pierces my sense of subjective existence (Sartre 2003, p. 287).[10] 'My being for others is a fall through absolute emptiness towards objectivity [. . .] myself as object [. . .] is an uneasiness, a lived wrenching away from the ekstatic unity of the for-itself, a limit which I can not reach and which yet I am' (2003, pp. 298–9). And Leder writes, 'a radical split is introduced between the body I live out and my object-body, now defined and delimited by a foreign gaze' (1990, p. 96). On this view, social existence of its very nature disrupts the transparent, effaced status of the body.

But even in everyday experiences where objectification is not a primary mode there are many ways in which the world resists us. Often the interaction between us and the world is smooth and automatic and regulated by mature behavioural repertoires. In these cases there is little

---

[10] But see also Merleau-Ponty's criticism of Sartre's analysis (2012, p. 378).

need for conscious attention of the body. But even in health the world may resist this smooth articulation and require conscious awareness. For those who, like me, have little DIY talent, the inability to glue together a broken vase or hang a picture straight, and the effort and clumsiness involved in trying, presents an experience of bodily limitation. The small knocks and resistances that we encounter in little accidents, bodily failures, bodily needs, and the inability to easily learn new bodily skills, disrupt bodily transparency in minor ways. However, these experiences are contained within a normal everyday, and are experienced on a spectrum of familiar, if frustrating, bodily failures.

Illness, in contrast, creates areas of dramatic resistance in the exchange between body and environment, so is wholly different to these small failures. Even if the transparency of the healthy body is somewhat exaggerated, and that transparency is frequently disturbed by social interactions and bodily failure, it is still the case that we intend towards the world through our body and it serves as a medium through which we encounter the world while remaining in the background. The body 'plays a constitutive role in experience precisely by grounding, making possible, and yet remaining peripheral in the horizons of our conceptual aware-ness' (Carman 1999, p. 208). Or as Merleau-Ponty said, the body is 'our general means of having a world' (2012, p. 147).

There are two ways of thinking about the healthy transparent body. One considers it as vulnerable as the conspicuous ill body; on this view the two are on a continuum. It is simply a matter of degree of vulnerability. Or we might think about the healthy and ill body as discontinuous: health and illness are distinctive bodily states, in which modes of being and experi-ence differ radically. I suggest that overall the discontinuity view is the stronger one. Although everyday experiences certainly include occasions when the body is explicitly thematized, and thematized negatively, these experiences do not fundamentally modify one's tacit sense of trust in one's body or disrupt the habitual body. Small injuries and bodily failures are experienced within a context of confidence and regularity, and thus are experienced as benign, even if frustrating or painful.

In contrast, the ill body, which becomes conspicuous like Heidegger's broken tool, takes over one's way of being, constricting the range of possible actions and hence restricting choice. It also constrains actions that are chosen from within that reduced range. The actions of the healthy body enable projects, while those of the ill body disable or delimit. For example, if a healthy person goes sightseeing in London, they will experience hunger and fatigue. But the project—seeing London—will not be hampered or shaped by these needs. For an ill person (e.g. a wheelchair user or someone who requires oxygen) the entire possibility of sightseeing is conceived within the constraints of illness (which buildings have wheelchair access? How much oxygen can I carry with me?). Thus they are not experienced as free projects, but as projects delimited by such restrictions. The sense of freedom, possibility, and ease with which the transparent body operates is entirely different to the limitations and anxieties of the conspicuous body. The constraints taint the freedom and spontaneity which underlie projects.

However, one may argue, even a minor headache can bring to light the tacit sense in which all projects ultimately rest on bodily abilities (Sartre 2003; Solomon personal communication). Minor ailments are also philosophically revealing. But there is a fundamental difference between a headache and the serious illness that is the topic of this book. Minor ailments fit within, and hence do not disrupt, one's being in the world. Serious illness modifies the ill person's way of being. A headache will make my head conspicuous, and will be experienced as the frustration of an action, but it will not permanently and radically modify my bodily experiences and self-understanding in the way serious illness does.

Let us look at Sartre's example. I am reading a book and while doing so the body is given only implicitly. Then my eyes start hurting. The pain is not perceived separately to the project of reading. Rather:

> this pain can itself be *indicated* by objects of the world; i.e., by the book that I read. It is with more difficulty that the words are detached from the undifferentiated ground which they constitute [. . .] consciousness exists its pain [. . .] pain *is precisely the eyes* in so far as consciousness 'exists them' (2003, p. 356). [. . .] pain in the eyes is *precisely my reading* (p. 358).

As Sartre points out, when the pain recedes, it disappears (p. 360). Minor illness rises and then subsides without fundamentally altering the structures of experience. But serious illness causes a fundamental change to one's embodiment, habits, ability to plan and pursue goals, and sense of freedom (this will be developed further in the next three chapters).

Although the two modes of being—the transparent and the conspicuous body—stand in contrast, they still mutually imply one another, as Leder notes. The appearance of the ill body (what Leder calls 'dys-appearance') is made possible because of the disappearance, or absence, of the healthy body. As Leder writes, 'it is precisely because the normal and healthy body largely disappears that direct experience of the body is skewed toward times of dysfunction. These phenomeno-logical modes are mutually implicatory [. . .]' (1990, p. 86). The contrast drawn here is not intended to deny that there are neutral and positive ways in which my body appears to me in health. The experience of dancing in front of an audience, for example, may be one in which pleasure is gleaned from the explicit thematization of the per-forming body.

In such cases pre-reflective experience may be accompanied by an explicit appreciation of the feeling of pleasure. Illness experiences are qualitatively different as they appear as a demand. The body does not appear simply to note its pleasurable state; it appears with a sense of urgency and a demand to do something about the pain, discomfort, or nausea through which the body comes to the fore. There is a heightening of bodily focus at times of illness and disruption. 'It would be a mistake to equate all modes of bodily thematization with dys-appearance', Leder notes (1990, p. 91). I suggest that it would equally be a mistake to think that positive and negative ways of appearance have more in common than the explicit thematization of the body.

2.5   The Body as Tool

Heidegger's tool analysis is based on his distinction between present-at-hand entities (*vorhanden*) and ready-to-hand or 'handy' entities (*zuhanden*) (1962, p. 96ff.). On this analysis, we do not perceive entities as mere objects, but as tools with which we go about our daily business (toothbrush, writing pad, frying pan) or set about pursuing our projects. Under normal circumstances, objects are never mere things, but handy tools and pieces of equipment which together form 'equipmental total-ities', such as offices (made up of desks, chairs, computers, reading lamps, telephones, etc.), farms, or hospitals.

These are not just spaces in which objects are arranged, but a mean-ingful practical totality of tools that makes sense to us *qua* functionally interrelated tools. Our knowledge of ready-to-hand entities is not abstract. Rather, 'the hammering does not simply have knowledge about the hammer's character as equipment, but it has appropriated

this equipment in a way which could not possibly be more suitable'
(Heidegger 1962, p. 98). Practical activity is not epistemically blind but
'has its own kind of sight', says Heidegger (p. 98). Tools are characterized
by their inconspicuous presence. When I use a pen to write a letter, it is
my friend and what I want to say that are at the forefront of my mind.
The pen is a tool that recedes to the background and is effaced by its
function, swallowed up by the task at hand.

It is only when such a tool breaks down that it becomes conspicuous.
Heidegger distinguishes three modes of dysfunction: *conspicuousness*—
the tool has ceased to work (the bike has a puncture); *obstrusiveness*—the
required tool is missing ('where are those keys?'); and *obstinacy*—the tool
is there but is unsuitable ('not *those* keys, the house keys!') (Heidegger
1962, pp. 102–3). Conspicuousness arises because the ready-to-hand
entity is 'unready-to-hand'; it has become present-at-hand. The 'unusa-
bility' of a tool is not discovered theoretically, but by trying, and failing,
to use it. 'The helpless way in which we stand before it is a deficient
mode of concern [practical engagement], and as such it uncovers the
Being-just-present-at-hand-and-no-more of something ready-to-hand'
(pp. 102–3).

This taxonomy of tool breakdown invites the analogy to illness. We
can think of the inconspicuousness that characterizes the functional tool
as also characterizing the healthy body. When my body does what I want
it to do (keep my balance when I am walking, digest food I've eaten) I pay
little attention to it and often have no knowledge of the biological
mechanisms involved. I live in (or as) my body and experience the
world through it; much of the time my attention is directed away from
the body to the object or task I am engaged in. In the case of the pen,
attention is deflected from the pen to the contents of the writing. So the
analogy with Heidegger's tool breakdown analysis seems to hold. Now let
us turn to the hand holding the pen. Imagine that the pen works perfectly
but I cannot use my hand—it is paralyzed ('conspicuous'), or amputated
('obtrusive'), or I have had a stroke and no longer remember how to write
('obstinate'). In these cases, too, I experience a failure of a tool, but this
time the tool is part of my body.

The duality of the body as object comes into play here. Viewed as a
physiological machine, we can indeed think of the hand as a malfunctioning tool and use Heidegger's tool analysis to capture such cases. But if we
think of the body as experienced and lived, its failure will be experienced

differently to the failure of the pen. Whereas we can throw out the useless pen and buy another, our hands (and bodies) stand in a fundamentally different relation to us. Our bodies cannot be replaced or repaired as readily as tools. Bodily dysfunction is experienced deeply differently to tool breakdown. My head with a headache remains attached to me and becomes increasingly conspicuous, increasingly disabling. Moreover, my head with a headache is not a malfunctioning tool, but a way of being. My head as pain means that I experience my head as a region of pain, not as a malfunctioning brain.

Such dysfunction also affects how we inhabit the world. My head with a headache is experienced as the frustration of the attempt to read (Sartre 2003), as a darkening of my existence, as a demand for me to address the pain. So although under a certain objective mode the body is a tool, it is also our medium for having a world. As Sartre observes: 'when by means of universalizing thought I tried to think of my body emptily as a pure instrument in the midst of the world, the immediate result was the collapse of the world as such' (2003, pp. 363–4). In other words, the body is not a tool in a crucial sense: it is the origin of our sense of being in the world and this feeling of inhabiting a world, although tacit, is anchored in the body and depends upon it (compare Ratcliffe 2013).

Because of this fundamental dependency of the sense of reality (of having a world that is familiar and feels real) on the body, a change to the body leads to a change in the experience of the world. We experience the world not only empirically, as a source of data, but also existentially and emotively. If a person goes blind, for example, she no longer experiences the world visually, and loses information from visual stimuli. But her existence as a blind person will affect other senses, her confidence, her projects and preferences, and so on. Therefore the change in the case of bodily dysfunction runs much deeper than tool breakdown, despite the fact that a similar process of becoming conspicuous characterizes both tool and bodily breakdown. The possibility of bodily breakdown is a fundamental yet unacknowledged aspect of our being and our experience of the world. Illness is a painful and violent way of revealing the intimately bodily nature of our being.

Bodily breakdown need not be dramatic in order to reveal our vulnerable nature to us. We have seen that even simple bodily disruptions, like a headache, may still reveal to us the contingency of our bodily being, although they do not modify the structure of experience in the way that deep disruption does. The headache may not be severe and it is transient. But even a simple headache disrupts the activity one is immersed in and thus reveals how our immersion in the everyday world is dependent upon bodily integrity. Vulnerability, limitation, and finitude are fundamental features of human life not only in its physiological objective mode, or as abstract knowledge, but also in its experienced, subjective mode, as informing our ways of being in the world.

2.6 Conclusion

This chapter opened by examining Toombs's essential features of illness as five losses: loss of wholeness, certainty, control, freedom, and loss of the familiar world. These were discussed and some ways in which Toombs's view can be expanded were outlined. We then looked at the distinction between the objective body and the body as lived, that Husserl and Merleau-Ponty put forward as a fundamental view of the body. We saw how this distinction is useful for understanding the difference between disease and illness. This account was then supplanted by Sartre's three orders of the body: the body as objective and as subjective, but also the order of intersubjectivity, or of the body as I experience it as reflected in the experience of it by others. We discussed the third order of the body as prominent in diverse experiences that make us conscious of how our body is perceived by others and how this affects our understanding of our body. The three orders were then discussed in relation to interactions between health professionals and patients.

I then presented the view that body is transparent, or absent, and asked whether it really is so, offering examples of lost transparency in health. I suggested that a deep qualitative difference separates health and illness (as it is defined in this book, i.e. serious illness) and that this is not only a difference in the contents of experience but in one's being-in-the-world and in the structure of experience. I concluded by examining the analogy between bodily and tool breakdown, using Heidegger's tool analysis, arguing that although the body is not a tool the analogy still holds and sheds light on illness.

These five phenomenological analyses cumulatively make up a framework for understanding the experience of illness, underpinned by the central role the body plays in experience. Any substantial bodily change results in modification not only of the contents of experience, but also its structure and conditions of possibility. Because illness changes the body, and the body is central to all experience, illness changes the structure of experience, as well as its content. This framework makes this impact both salient and effable. Illness is not merely a suboptimal dysfunction of a body subsystem (compare Boorse 1977), but a systematic transformation of how the body experiences, responds, and performs tasks as a whole. The change in illness is not local but global; it is not external, but at the core of the self.

3

The Body in Illness

> The body when ill is a concert master not only of pain but of warmth and cold, bloating, pressures, fatigues, nausea, tinglings, itches.
>
> Drew Leder, *The Absent Body*

This chapter uses the framework developed in the previous chapter to examine how the lived body changes in illness, as well as how the physical and social world of the ill person are modified. The first section returns to Toombs's characterization of illness as a series of losses. The second section discusses illness as bodily transformation. The third section examines the social architecture of illness and the fourth section characterizes illness as 'dis-ability'; as being unable to be or do that is restrictive but also profoundly human. Throughout the chapter I also touch on the difference between chronic and acute illness, somatic and mental disorder, congenital or acquired illness, asking whether the descriptions and analyses offered here capture all, or some, of these categories.

3.1 Illness as Loss

Let us begin by returning to Toombs's (1987) five losses, which she suggests characterize any kind of illness: loss of wholeness, certainty, control, freedom, and familiarity. First we should ask if these losses would characterize all types of illness: chronic and acute illness, somatic and mental disorder, congenital or newly diagnosed illness. It seems that bar a few exceptions, Toombs's losses do indeed capture a fundamental experience of illness, in which the ill person feels that something is taken away from her, something which falls under these five broad types of losses, as discussed in the previous chapter.

When considering the broad category of illness we need to distinguish between cases where the ill person expects the illness to be temporary, and is confident that what has been lost will be regained, and cases where the loss is uncertain, enduring, permanent or progressive.

Sometimes illness can lead to a sense that something has been lost, while in other cases the ill person may feel only temporarily obstructed by the illness. Any general analysis of illness needs to be sensitive to the mode of loss experienced by the ill person. A loss is experienced differently if it is considered temporary (e.g. fractured bone), permanent (e.g. amputation), or progressive (e.g. aggressive cancer). It may become particularly oppressive and difficult to endure if it is uncertain (progressive disease where the rate of progression is variable or unknown).

However, if we look outside modern Western culture, we find that illness experiences are interpreted differently. For example, the twelfth-century nun Hildegard of Bingen suffered from migraines with visual disturbance, which she experienced as religious visions (Sweet 2006). Modern understanding of the Oracle of Delphi's prophecies relates her divine inspiration to the inhaling of toxic fumes.[1] In other cultures conceptions of the body and of illness include a deep spiritual element (Yoeli-Tlalim 2010). So illness, which is distinctly aligned with loss on Toombs's account, might be perceived differently in other cultures. In addition, our culture privileges youth and health and perceives illness as a form of weakness or even personal failure (Ehrenreich 2010; Frank 2010). Other cultures and periods value old age and the wisdom and experience associated with it, so may have a very different attitude to illness and frailty. Cicero, for examples, describes old age as 'the tranquil and serene evening of a life' in which one can concentrate on 'the study, and the practice, of decent, enlightened living' (1971, pp. 216–18).

Although Toombs is right that illness involves loss, it can also involve gains. Such gains might include personal or spiritual insight, moral maturity, edification, and emotional clarity, as well as an ability

---

[1] See <http://news.nationalgeographic.com/news/2001/08/0814_delphioracle.html> accessed 4 October 2015.

to focus on what is most important and let go of other things (see Chapters 6 and 9). In some cultures, it is necessary to undergo losses such as those Toombs describes in order to become capable, or even deserving, of such gains. Crucially, the possibility and intelligibility of such patterns of loss and gain depend on a wider conception of the nature and value of illness and of human life. And these, of course, vary greatly both across cultures and historical periods. Our starting point is therefore to take Toombs's analysis as part of a phenomenology of illness, but qualify it as applying to Western contemporary culture, rather than revealing essential features of experience, as Toombs suggests.

How are these losses experienced in illness? Let us think about the course of illness, from appearance of symptoms, to diagnosis, disease progression, and prognosis. These phases are not consecutive and may overlap or appear in cyclical form, in the case of repeated exacerbations followed by temporary or partial recovery (e.g. in asthma, allergies, multiple sclerosis). The description that follows is therefore not intended as a series of discrete phases but as overlapping, often parallel, aspects of illness that unfold in different ways and rhythms.

Symptoms normally precede a visit to the doctor, sometimes by months or years, although on some occasions diagnosis is made prior to experienced symptoms, for example, if abnormality is picked up in routine screening. Let us take the example of respiratory illness. By the time respiratory patients become aware of symptoms and consult a health professional they may have lost 20 to 50 per cent of their lung function (GOLD 2010). One explanation for this apparent anomaly is that many people do not exert themselves sufficiently in daily living to become aware of increased breathlessness; as a result, by the time a diagnosis is made, the damage to lung tissue and loss of function are usually moderate to severe (GOLD 2010). Symptom appearance and the period before a diagnosis is made can be one of increasing anxiety and sense of abnormality as well as decreased ability. Even prior to diagnosis, symptoms can be experienced as a loss of freedom and certainty.

Another common loss experienced when new symptoms appear is loss of control. Symptoms such as incontinence, fainting, and vomiting are uncontrollable and may cause extreme embarrassment and grief because they enact a loss of control over bodily functions. But less dramatic symptoms, such as muscular weakness, fatigue, or mild pain, can also give rise to a sense of loss of control. 'What is happening to me?' is a common reaction to a new negative sensation such as pain, breathlessness, or fatigue. Familiar bodily sensations are replaced by alien, negative, bodily feelings that are experienced as loss of control. This loss is mirrored by the loss of familiarity and wholeness of one's body. Bodily integrity may be suspended or permanently lost when new symptoms appear. Of all of these losses, I suggest that the loss of freedom, in the expanded sense I offered in Chapter 2, is the most prominent.

Illness is the loss of opportunities, possibilities, and openness. It is the closure of a previously open future: future possibilities close down as illness progresses. But it is also closure of the present: current daily activities lose their habitual aspect and become carefully planned and demanding. What could once be done unthinkingly, with no planning and marginal effort, is now an explicit task, requiring thought, attention, and a pronounced effort. This shift takes place when symptoms first appear. Indeed, the time of symptom appearance, prior to diagnosis but also after, is a time of great change, upsetting previous life habits. Mundane actions such as running for the bus may move to the realm of fantasy. They are no longer live possibilities.

Diagnosis turns symptoms into a less subjective entity. They are now organized in an explanatory pattern that excuses, explains, and predicts illness behaviour. In this sense diagnosis can be experienced as affirmation of subjectively experienced symptoms, making one not 'just a complainer' but someone who has a genuine medical condition justifying certain adjustments. For example, many women presenting with breathlessness due to the respiratory disease lymphangioleiomyomatosis (LAM) are diagnosed as having an anxiety disorder, panic attacks, or another mental disorder. When the correct diagnosis is made, the patient may feel vindicated, that her complaints which may have been interpreted as merely subjective sensations are a 'real disease'. In addition, there may be psychological and social gains to be made by a diagnosis. One is no longer a 'moaner' or hypochondriac, but someone with an official and ontologically robust diagnosis. However, such utility must be tempered by the loss of subjectivity: the diagnosis also signals an appropriation of my pain, my 'stomach as painful', by the other's point of view. Sartre writes:

At this point a new layer of existence appears: we have surpassed the lived pain toward the suffered illness; now we surpass the illness toward the *Disease* [. . .] It is then objectively discernible *for Others*. Others have informed me of it, Others can diagnose it; it is present for Others even though I am not conscious of it. Its true nature is therefore pure and simple *being-for-others* (2003, pp. 379–80).

The time of diagnosis is the time in which the illness becomes known by others and by the ill person as disease. It becomes objective (and often objectified) and subjected to medical management, labelling, and so on. This movement from a private, subjective experience to an objectified disease, which continues to be experienced as symptoms by the ill person, is a significant transition. The illness is no longer a private musing on the nature of an unexpected bodily change, but an item in a medical vocabulary and ontology, to which shared meanings and knowledge are attached.

A patient's hospital file, pushed around on a little trolley, exemplifies the appropriation of illness by disease. The file contains test results, letters to and from specialists, and requests for further tests, but nothing else. It is a file about the patient, but not with her. In my case, that bulging file with its stapled addenda symbolizes the subsuming of breathlessness, pain, suffering, social awkwardness, bodily failure, and fear of death, under a medical gaze. And under that gaze the lived correlates of the medical information are often relegated to the subjective, and hence secondary, pile.

Disease progression is probably the phase at which losses are experienced most acutely. The continuous denigration of freedom is experienced as diminished bodily capacities or increased reliance on medical aids, but also as a deepening erosion of one's freedom to plan and live. What may be lost is not only one's freedom to plan and live but also one's desire to do so—a loss, perhaps, of positive affective response: if one lacks the desire to act, then an actual incapacity to act becomes, in a sense, secondary. Disease progression is the most intense enactment of our finitude, of our realization not only of mortality but also of our bodily vulnerability and interdependence (Carel 2013a; MacIntyre 1999). Here is the description given by the ill Ivan in Tolstoy's novella *The Death of Ivan Ilyich*:

That Caius—man in the abstract—was mortal, was perfectly correct; but he was not Caius, nor man in the abstract: he had always been a creature quite, quite different from all others [. . .] And Caius was certainly mortal, and it was right for him to die; but for me, little Vanya, Ivan Ilyich, with all my thoughts and emotions—it's a different matter altogether. It cannot be that I ought to die. That would be too terrible (Tolstoy 1995, p. 54).

Disease progression is frightening because it curtails possibilities but also because it may be part of dying. As Ilyich's illness progresses, his bodily experiences of pain, exhaustion, and helplessness take prominence. Later on in the novella he is described thus: 'He waited only until Gerassim had gone into the next room and then restrained himself no longer but wept like a child. He wept at his own helplessness, at the cruelty of man, the cruelty of God, at the absence of God' (Tolstoy 1995, p. 76). It is not only the fact of his death that debilitates Ilyich. It is the sense that his having lived his life as an autonomous, self-sufficient, and independent man is peeled away in his dying.

His surrender to his dependence on Gerrasim, his servant, and on his doctor to supply him with morphine to alleviate his pain, and his surrender to his own death, is Ilyich's ultimate transformation. He heeds MacIntyre's call for us to acknowledge not only our vulnerabilities and affliction, but also our consequent dependence on others, advocating 'the virtues of acknowledged dependency' (MacIntyre 1999, p. 8). The illusion of autonomy and independence, and the misunderstanding of adulthood as subsuming the whole of human life, are two errors that lead to an inadequate moral view, argues MacIntyre. Ilyich's view is transformed through his illness: from being solely interested in doing things *comme il faut*, to authentic conversion. Ilyich's self-understanding and his struggle to resist dependence are given up at the end of the story, replaced by acceptance.

It occurred to him that what had appeared utterly impossible before—that he had not lived life as he should have done—might after all be true [. . .] And his professional duties, and his ordering of his life, and his family, and all his social and official interests might all have been false. He tried to defend it all to himself. And suddenly he realized the weakness of what he was defending. There was nothing to defend (Tolstoy 1995, pp. 83–4).

What Ilyich learns through his gradual decline and movement towards death are dependence and humility. He also experiences the losses we started out with: loss of wholeness, certainty, control, freedom, and familiarity. By the end of his life, everything is lost. But Ilyich also learns that life is fragile and precious; that satisfying social expectations amounts to very little; that he lacks real intimacy.

Illness affects one's entire way of being; the narrow medical view of illness cannot capture these dramatic and intimate changes to one's life and being. A phenomenological approach to illness can illuminate these changes and thus helpfully augment clinical medicine. In the following sections I discuss these changes, starting from the physical world of illness. I then turn to social and psychological changes, which include changes to identity, self-perception, and emotional well-being.

Later on, in Chapter 7 I discuss finitude, temporality, and death. These are major philosophical topics that have been studied for millennia. However, the effects of illness on personal identity, well-being, and temporality have been given little attention in the philosophical litera-ture. Thus this examination is important not only for our understanding of illness, but also for our views about philosophy. Returning to the bilateral relationship between illness and philosophy discussed in the Introduction, these sections play a double role: they are both an explor-ation of the impact of illness on the ill person's life world and preparatory work for the final chapter of this book, which explores the philosophical role of illness. A final comment about this chapter: although it is divided into sections, this is not to suggest that the different life domains are discrete, or that changes in one domain do not also imply changes in other life domains, as will be demonstrated below.

3.2 The Transformed Body

In illness things grow heavier and further away. A distance I would once call 'near' or 'a day's walk in the countryside' is now 'far' or 'impossible'. Small tasks like carrying groceries home or lifting a child require prep-aration, pauses, rest, and cause fatigue. Everything is hard. Everything is far. Everything is strenuous. My world, and the world of those who are close to me, has shrunk. For me, the trap is permanent. There is no release from it. Every movement requires oxygen. This fundamental fact about human biology, known to us all in the abstract, is experienced in everything I do.

In respiratory illness, the limitation is felt continuously. There is no respite from the exertion and breathlessness that accompany almost everything I do. The aim is to avoid the crazy intense panting, the tightening chest, the doubt whether I will be able to catch my breath, and the knowledge that there will come a moment when I will no longer be able to do this. 'Respiratory failure' is the medical term; failure indeed. My failure, my lungs' failure, failure in gas exchange, in inhalation, in loading red blood cells with precious oxygen, in exhalation, in emptying the lungs of air in order to refill them with yet more air, more oxygen. Being perpetually breathless is, more than anything, deeply uncomfortable. Movements are censored; activities struck off the list of possible ones. Slow, regal, paces; slow small steps to go uphill. Bigger, but still slow, paces to go downhill. Uphill: walk, pant, walk, pant pant, walk more slowly, pant pant pant pant, stop. Rest, catch breath, walk some more, pant some more. Every step and every breath fill my consciousness, control my mind, leave no space for anything else.

When I am upbeat, I point out that I get to stop and smell the roses. But normally I am scanning street signs, staring at dormant winter foliage, or counting pavement stones. If someone walks past, I sometimes check my phone, to look less aimless and weird, standing in the middle of a street in the pouring rain, while everyone else runs for cover. If someone tries to walk with me, I am full of dread. They start out in their normal pace, then slow down to wait for me, then slow down more, realizing how slowly I walk. Now they, too, are conscious of their walking, of their breathing, maybe of their wellness. They are bored so they try to converse: bad move. I'm too breathless to talk. We stand in awkward silence; unspoken pity fills the space between us. They move to monologue to cover over the pity and the awkwardness, but I can neither listen nor engage. I haven't enough air to think, let alone talk; my breathing takes up all my effort and concentration. Eventually, we walk in silence. Me: bitter, exposed, ashamed. People eventually learn not to walk with me. Walking has become for me an intense and private affair.

I now suspiciously scout important places before meetings or gatherings. Where can I park? Where is the toilet? Is the restaurant seating upstairs? I sometimes send my husband and son on a reconnaissance mission. They go first, check out the place, or the walking route, or the cycle track (I have an electric bike) and I then follow. I envy them. They can just veer to the right or left, leave that flat riverside path, and climb up into the woods. They have the freedom to stray from a planned route. They can scale a mountain, climb up, run down. They don't need to ration their energy or conserve oxygen. They enjoy an unencumbered physicality, and the freedom it brings with it. I am trapped by routine, by pre-planned routes and timings. Spontaneity is lost.

This kind of loss of spontaneity characterizes the case of Schneider, the First World War soldier, who suffered a head injury causing him to lose the ability to plan, think abstractly, and fantasize. Merleau-Ponty (2012) describes Schneider's malaise as 'existential': he has lost the capacity for spontaneity, intellectual creativity, and playfulness. Schneider is '"bound" to the actual, and he "lacks freedom", he lacks the concrete freedom that consists in the general power of placing oneself in a situation' (p. 137). While Schneider lost the existential capacity for spontaneity because of a brain injury, I have lost my capacity for spontaneity because everything needs to be pre-planned. I have to stick to the planned path, and the planned time, lest I run out of oxygen. Day excursions are limited to three-hour chunks, after which I must return to base to refill my oxygen cylinders. Aimless strolling is impossible and any outing is accompanied by a consciousness of time lapsing, oxygen being used up, a cylinder gauge going below the halfway point. Joyful ambling in the countryside or an excursion to a new city become stressful and circumscribed by the need for oxygen and rest. Schneider cannot imagine, and therefore cannot execute. I cannot execute but can easily imagine. Thus illness can destroy creativity in one of two ways: either by removing the capacity to fantasize or by removing the capacity to execute.

The ill body is transformed in a number of ways: it is now experienced through the losses Toombs describes (1987) and as discussed above. In addition, the ill body is experienced explicitly, and often negatively, rather than transparently. The naturalness, if not transparency, that characterizes normal bodily commerce with the world is replaced with artificial and explicit attention to the body. This attention may be related to a medical assessment of the body: 'has the cancer progressed?'. It is also related to the everyday execution of routine tasks. Explicitness with respect to movement, effort, bodily functions, fatigue, and so on, is often a part of illness. For example, a diabetic must assess before a meal how much and what they intend to eat and drink. They then need to calculate how much insulin to inject. And they then need to stick to the calculated amount, so spontaneity is lost.

The natural way in which we may sample a new kind of chocolate in a tasting stall, or pour ourselves a glass of orange juice becomes a carefully managed affair. The body is also experienced more frequently as an object of medicine, and may be further objectified with cumulative exposure to medical examination and treatment. When looking at test results of kidney function, or the images from a CT scan or X-ray, one sees one's body as never before. The invisible interior becomes visible, available for one's own scrutiny; an anxious anticipation of the medical pronouncement on one's kidney function or size of tumour arises.

The gap between the lived and objective body is experienced more acutely and this punctuates the ordinary feeling of embodiment. In addition the three orders of the body are more prominently experienced as different because of feelings of shame and self-consciousness that often exacerbate the misery of illness. The body is no longer transparent. Bodily breakdown becomes a common experience (see Chapter 2, this volume). As Merleau-Ponty notes:

> [. . .] the procedures that [illness] employs in order to replace the normal functions that have been destroyed are themselves pathological phenomena. The normal cannot be deduced from the pathological, and deficiencies cannot be deduced from their substitutions, through a mere change of sign. The substitutions must be understood as substitutions, as allusions to a fundamental function that they attempt to replace, but of which they do not give us the direct image (2012, p. 110).

These disturbances characterize, I suggest, all kinds of illness: chronic and acute, somatic and mental, congenital or acquired—all the disorders subsumed under these categories give rise to a change in one's body and world. Even mental disorder, which may seem to affect the mind rather than the body, reveals, when studied phenomenologically, substantial changes to one's sense of embodiment, bodily possibilities, and bodily feelings (Ratcliffe 2008a, 2012a; Stanghellini 2004). Even if such changes are not experienced distinctly as a loss, they still characterize illness in the broadest sense: illness is a profound alteration of one's bodily experience.[2]

3.3 The Social Architecture of Illness

How do you introduce yourself to people if you have a serious health condition? What do you say? How can you assuage their discomfort, the sense that you are an alien being, with your wheelchair, or insulin

---

[2] The case of congenital illness is different. Here I suggest that the sense of change is related to the gradual realization of one's different needs and capacities.

injections, or oxygen tubes and all their associated complexities? How can you carry on being socially 'normal' when illness shapes nearly everything you do? How do you handle chance encounters with old acquaintances who don't know about your condition? Where do you begin?

These are not medical issues, but they certainly shape the experience of illness. The change in embodiment and self-perception discussed earlier is mirrored by a change to social perception. How the ill person is perceived by strangers, friends, and acquaintances will shape her illness experience. Stigmatization can be incredibly costly for the stigmatized individual in terms of social relations, but also job prospects, income, and support networks. It is particularly acute in the case of mental disorder, even a common one such as depression (Blease 2012). This section also considers the role of friendship and the strains placed on it by illness. The experience of bodily betrayal and disappointment, the threat illness poses to intimacy, and fear of the diseased body all impact on our relationships.

The transformation is most visible and damaging in the ways it hampers the ill person's social participation, narrows the range of available activities, and makes interactions difficult. The ill person might be unable to participate and reciprocate in social events (e.g. inviting people to dinner, if cooking is difficult), there may be awkwardness around the subject of illness or disability, the ill person may fall out of step with others' activities and interests, and she may enter into a vicious cycle of increasing isolation and depression. These alone could cause severe damage to a person's social world. But there are other problems: it is difficult to ask for or know when to offer help, people often experience unease around conversations about illness, and harshest of all, friends may stay away because they do not know what to say (Carel 2013a).

Visible illness or disability often becomes the elephant in the room, unless the ill or disabled person actively leads that elephant out of it. Illness is often seen as something that is not to be mentioned by polite people, who mustn't draw attention to the illness. But at the same time the medical condition challenges normal interactions and makes 'not talking about it' difficult, sometimes impossible. People often feel they ought to say something, but are not sure what to say, or how, or when. They feel they should censor their expressions and self-reports, so as not to offend the ill person, but also feel curious, or disgusted, or admiring towards her. The result is a general sense of discomfort, being ill at ease and unable to transcend the social barrier created by the illness (Carel 2013a). In addition, not only are existing relationships changed or lost, but one's expectations about the possibility of forming new relationships might also be negatively affected.

Illness and its visible signs may arouse strong emotional responses in healthy onlookers or friends. These emotions may not be consciously experienced and cannot be addressed in a routine exchange. It is difficult to find the right time and words to express these feelings. I witnessed many attempts by people to offer encouragement and support, to express admiration and caring towards ill people. The striking feature of these attempts was how difficult they seemed for the well-intentioned healthy person. They often preface their comments with an apology for being intrusive or speaking out of turn.

There are additional problems facing an ill or disabled person in their social interactions. There are practical problems, such as being unable to participate in some social events such as going for a walk, going dancing, or drinking. Everyday activities have to be modified or sometimes given up if the condition does not enable the ill person to take part in them. The ill person can feel she is slowing the others down or hampering the natural flow of events merely by being present. This, in turn, leads her to give up attending these events and a vicious cycle may begin. There are also novel social issues that arise from the illness. For example, the ill person may feel apprehensive about meeting new people because of the awkwardness created by the illness. She may feel the need to explain her condition but also reluctance to do so. She could feel nervous about leaving the house and going to unknown territory, where the number of steps or wheelchair access is unknown. She might not have the energy to participate in some activities, or fear that it would take too much effort for her, or that she will embarrass herself by not being able to keep up.

This social architecture of illness mirrors the transformed body discussed in the previous section. In the same way that distances increase, hills become impossible, and simple tasks become titanic, the freedom to go out into the social world and improvise, to act and interact, is similarly reduced. A new world is created, a world without spontaneity, a world of limitation and fear: a slow, encumbered world to which the ill person must adapt. Many people experience this loss of spontaneity through ageing. In illness this opaque and alien world can emerge overnight.

This is a world of negotiation, of helplessness, of avoidance. It is an encounter between a body limited by illness and an environment oblivious to such bodies. Schneider's existential malaise is mirrored here. Being able to stay up for a night out, to dance, or just walk from one place to another, being able to converse and laugh, to have enough energy for socializing, to be able to stand up talking to people for long periods, all of these abilities might be gone, or damaged. The spontaneity that enables social relations to develop may be lost or damaged. In other words, the way that the physical and social environments intersect can make it even harder to negotiate while ill.

Ill and disabled people invent a myriad of strategies and coping mechanisms to override the constraints inflicted on them by the environment and by the invisible background norms that govern our lives. The demand to be autonomous, independent, self-sufficient, is often met by failure in cases of illness or disability. And perhaps if there wasn't such a premium on autonomy and independence, the damage to such people's social life and self-esteem would be lessened. That the premium is so high indicates that what might really trouble some people about illness is that it puts the lie to many inflated claims about autonomy and independence. Perhaps illness reveals to those who find it so hard to encounter it that they are not the autonomous independent beings they thought they were, but are dependent upon others and therefore need both trust and respect.

These problems lead some ill people to become less sociable and to participate in fewer social events than previously. If we return to the transparency of health discussed in the previous chapter, we can see that the transparency of health also includes a modicum of transparency of social existence. The natural way in which we engage in social interactions becomes cumbersome in illness, weighed down by unspoken doubts and discomfort, and the effort required for genuine communication becomes greater. The social impact of illness is the loss of this transparency and immediacy of social interaction.

This transparency of the body, of social ease, can be characterized more generally as a transparency of well-being. Well-being is the invisible context enabling us to pursue possibilities and engage in projects. It enables us to follow through aims and goals, to act on our desires, to become who we want to be. But the spatial and temporal possibilities that characterize health are altered in illness, as we saw in the previous chapter. This is not only the curtailment of spatial possibilities, but the abrupt descent of limits onto a world previously larger, freer, more open. These limits not only restrict physical movement, but inflect existential possibilities. It is not only physical possibility that suffers at the hands of illness. It is ways of being and ways of being-with.

3.4 Illness as Dis-ability

Illness, and in particular a poor prognosis, can have a deep psychological effect on the ill person and those around them. A distinctive feature of illness is a sense of helplessness, loss of control, and vulnerability. These stem from lost bodily capacities, and the disability and dependence that stem from this loss, but also from the inability to control the disease process from exacerbating or barring the ill person from doing certain things. The ability to care for oneself, but also the autonomy to make one's way in the world, is seen as a fundamental feature of adult human life. Although this view has been criticized (MacIntyre 1999), it remains a standard view in medical ethics (Varelius 2006; Casado da Rocha 2009). We will therefore use this as a starting point of this section's question, namely, what happens to one's life when one becomes restricted by illness?

Heidegger characterizes human existence as 'being able to be' (*Seinkön-nen*) (1962). For him, human existence is distinctive because of its openness, potential, ability to become this or that thing. This underpins an existentialist view: one can become what one wants by taking the relevant actions. If I want to be a polar explorer, I have to train, build up my strength, learn to navigate, and so on. Eventually, I could join a polar expedition and fulfil my goal. Of course the obvious physical and temporal limitations would apply, and I would be restricted by my 'thrownness' (*Geworfenheit*)—being born into a particular culture, historical period, etc. As Dreyfus notes, I could not become a Samurai warrior, because that option is no longer available in the place and time I am thrown into (Dreyfus 1991). Although this openness is restricted by common-sense limitations, it still characterizes human existence as singular in its freedom, openness, and power of self-determination. Our plans and aims connect present actions (e.g. studying navigation) to a future view of ourselves as being able to be a particular thing (a polar explorer). Present actions have meaning in virtue of being part of a project that is forward-looking. I do something now in order to become something in the future.

I cannot become something in the future by merely thinking about it or wishing for it. I must take relevant concrete actions in order to become what I want to be. This definition of the human being is best understood through Heidegger's notion of projection. Projection means throwing oneself into a project, which connects the present with the future, and is also informed by the past (thrownness). Projection defines a human being's character and identity. If my project is that of being a teacher, I project myself accordingly by training to be a teacher, applying for teaching positions, and so on. This, claims Heidegger, is the essence of human existence: the ability to be this or another kind of person, to become something, even if this does not ensue from a conscious decision to be this kind of person engaged in this activity.

This view of the human being as becoming, as pursuing aims, as constantly moulding herself according to the project she pursues, is appealing in many ways. It credits us with the freedom—and responsibility—to shape ourselves and our lives in a way we find fulfilling: to transcend our present self with a future self that is more developed, more able. This progressive view of the person sees it as constantly growing and developing, in line with the temporal structure of Dasein (Heidegger 1962). Dasein 'is temporal in the very essence of its Being' (1962, p. 428).

Summarizing the temporal structure of Dasein (in its everyday existence), Heidegger defines Dasein as: 'Being-in-the-world which is falling and disclosed, thrown and projecting, and for which its ownmost ability-to-be [Seinkönnen] is an issue' (1962, p. 225, translation modified, italics removed). As Merleau-Ponty says, echoing Husserl and Heidegger, being in the world is not a matter of an 'I think' but an 'I can' (Merleau-Ponty 1962, p. 137). The active, goal-pursuing, able Dasein is Heidegger's model of a human being. Although the paradigmatic cases of 'being able to be' seem to be those of playing a certain social role (mother, head of a tribe, husband), or of pursuing a vocation (polar explorer, teacher), Heidegger intends to characterize the human way of self-interpretation that informs and orders our activities, rather than an explicit choice of goals and conscious life-planning activity (Dreyfus 1991, p. 95). As Heidegger says, 'Dasein has assigned itself to an "in-order-to", and it has done so in terms of an ability-to-be for the sake of which it itself is—one which it may have seized upon either explicitly or tacitly [. . .]' (Heidegger 1962, p. 119, translation modified).

In illness the general capacity to pursue projects is lost (given constraints on physical agency) as well as the capacity to pursue specific projects (e.g. being a mountaineer). Perhaps the projects that now remain available in one's trajectory are quite distant from one's values, interests, and hopes. It is therefore those fundamental aspects that need to be modified as well, rather than a mere pragmatic adjustment of projects already chosen. In illness, as well as in other situations of dependency, insufficiency, and incapacitation, understanding the human being as 'ability to be' does not seem as useful or descriptive. In fact, one's first and final years are usually periods in which one's 'ability to be' (in the Heideggerian sense) is restricted and dependent on the facilitation of others. It does not feature the capacity for choice–making in the broad, explicit sense (e.g. as in choosing what career to have, or whom to marry). When thinking about Heidegger's characterization of the human being as 'being able to be' we need to consider human life as a whole, including parts of life in which we are unable to do certain things, and eventually completely unable to be (in death). We begin and end in insufficiency and dependence.

Heidegger's definition seems to only capture a limited part of human life in at least three senses. First, it only captures the middle part of the trajectory of a human life, excluding infancy and aspects of childhood and old age. Second, it only captures the paradigmatic cases of healthy, autonomous adulthood, in which the ability to be is not hampered by disability or illness. Third, it overlooks the important ways in which our existence depends on other people and is saturated by a background sense of trust (Bernstein 2011). There are many ways in which we are unable to be, are only partially able to be, or in which being requires extreme unsustainable effort.

Let us take as an example the case of physical ability to be, say, that of being able to be an athlete. Someone may be (and hence is able to be) an athlete for many years, in an enjoyable and straightforward way. She exerts herself and suffers in training, and sometimes pushes herself right to the edge, but she does not lose bodily control, pass out, harm herself, or experience physical breakdown. In this case we can say that she is able to be an athlete. Eventually her body declines, she cannot run so fast or jump as high, and a point comes when she is no longer able to be an athlete. What happens at this juncture to her ability to be? To her self-understanding as an athlete? To her sense of skilful performance, bodily control, and physical joy? Is her ability to be simply eradicated? A closer inspection of the process of becoming an athlete and no longer being able to be an athlete shows that inability is implied by ability. Being able to be an athlete is rooted in an organism that starts out and ends as unable to walk, let alone run. Heidegger's discussion excludes both ends of this natural trajectory, and thus overlooks this important aspect of life, that of decline, inability, and failure to be.

When ill or ageing, we become unable to do some things, perform particular roles, and engage in certain activities. This poses a problem for Heidegger's definition because it excludes important parts of human life and common human situations. In some illnesses, especially mental and chronic illness, a person's ability to be, to exist, is radically changed and sometimes altogether curtailed. For example, in severe psychosis the possibility of having a project at all may become impossible. Similarly, in severe depression the possibility of any goal-oriented action, whether the goal is explicit or tacit, is lost (Ratcliffe 2011).

In less extreme cases of illness, certain projects and ways of being must be discarded and sometimes a replacement for these is difficult to find. Three questions arise. First, does Heidegger's account allow radically differing abilities to count as forms of human existence? Second, how flexible are human beings in modifying their projects and goals? And third, how much can we adjust our projects and plans in the face of ill health and how should we think of such adjustment?

I suggest that Heidegger's definition should not be understood too literally and that his characterization of existence as 'being able to be' can be modified in two ways that would enable it to encompass illness and disability. First, the notion of 'being able to be' can be broadened to include radically differing abilities. Secondly, 'inability to be' needs to be recognized as a way of being that is other to death, which Heidegger defines as the complete inability to be, or 'the possibility of the impossibility of any existence at all' (1962, p. 307, italics removed). Heidegger's definition can be made more inclusive if we think about 'being unable to be' as a form of existence that is worthwhile, challenging, and, most importantly, unavoidable. In order to achieve that we should interpret the notion of 'being able to be' as broadly as possible. It should include cases in which the smooth operation of the body, its assistance in carrying out plans and projects, is disrupted or removed. It should also include cases of prognostic uncertainty, or uncertainty about one's ability to pursue a future goal. And it should also include cases of failure that arises from psychological and social barriers.[3]

As happens commonly in illness, current projects may have to be abandoned and new projects created; these new projects must be thought of in light of limitations imposed by illness. Such new projects therefore arise within a restricted horizon and are thus different to cases of simple 'ability to be', where no unusual restrictions limit it. 'As long as it is, Dasein always has understood itself and always will understand itself in terms of possibilities', writes Heidegger (1962, p. 185). But possibilities and their concrete availability to a particular individual are distinctly shaped by gender, race, political situation, mental and physical disability, and so on. It is naïve to think that most individuals' possibilities are not shaped and restricted to an extent by aspects of thrownness. This much is said explicitly by Heidegger. But the step he does not take is that of reconfiguring the notion of 'ability to be' in light of these restrictions.

I suggest that radically differing abilities, and even ones in which possibilities are curtailed, count as abilities to be, even if the freedom is experienced within a context of limitation. Take a person in a wheelchair, someone with advanced lung cancer, a person with learning disabilities, a child with Down's syndrome—all of these are ways of being that differ in significant respects from the paradigm case of health. But they are nonetheless human ways of being that contain elements of 'ability to be' within a broader context of inability. Perhaps the outcome of applying Heidegger's notion of 'being able to be' to cases of illness and disability is an acknowledgement of the diverse ways in which it is possible to be and the ways in which human beings differ in abilities and possibilities.

The opposite of being able to be is not being able to be; but this presupposes that the two notions form a dichotomy. We can replace this dichotomy with a spectrum of abilities to be. There are other modes of being able to be that are excluded by this dichotomy. Being partially able to be, learning to be able to be, and rehabilitating an ability to be are

---

[3] This is not intended as an exhaustive list.

a few examples. The ability to be that characterizes human existence is territory to be experientially explored and developed, rather than delimited through this opposition. We can easily find positive examples of this. Stephen Hawking may have wanted to be a footballer or pianist, but because of his motor neuron disease (ALS) was unable to pursue these projects. Instead he had another project, being a physicist, and has become extremely successful at it. It is true that many projects that might have seemed attractive to him were closed off because of his illness. Even within a contracted horizon of possibilities, there is still an ability to be. We can also think of processes such as rehabilitation from drug use or after a car accident; learning to enjoy life after severe depression; being only partially able to walk, hear, see, or talk, and so on. None of these conform to Heidegger's definition, but if we expand his notion of 'being able to be' as proposed here, can accommodate such cases within his framework.

Furthermore, in cases of ageing, disease, or disability we need to acknowledge inability to be as a way of being. One way of thinking about ageing and illness, and indeed one conception of those as forms of decline, is as processes of coming to terms with being unable to be. As coming to think of one's existence as more reliant and less independent, more interlinked and less autonomous. The inability (or limited ability) to be and do is the flipside of Heidegger's account. For some individuals it is there throughout life, as in cases of chronic illness or disability. For all of us it is there as experiences of inability and failure both in the stages of infancy and childhood, and of ageing and decline, if we conceive of old age in that way.[4]

In fact, even the most 'able to be' adult life is inevitably sandwiched between these two types of inability: before and after the brief period of healthy adulthood. Inability and limitation are part and parcel of human life, just as ability and freedom are. By introducing the notion of 'being

---

[4] It is worth noting that the evaluative status of both childhood and ageing is not fixed, but determined by changing conceptions, typically implicit, of these stages. It's useful to distinguish between *gerontophobic* and *gerontophilic* conceptions of ageing, as decline, loss, failure, or as consummation and completion, respectively. Confucius and Seneca, for instance, were gerontophiles, regarding old age as a state in which the project of moral and civic achievement could be brought to completion, and hence as a state of satisfaction and quiet pleasure. They would be baffled by modern cults of youth, although of course we need to temper this enthusiasm with cases in which ageing does not bear such ripe fruit. I thank Ian James Kidd for bringing this point to my attention.

unable to be' as an integral part of human life, we can move from seeing ability as positive and desirable to seeing it as part of a broader, more varied flux of life.

Being unable to be is not an independent or context-free concept. It has to be seen in relation to being able to be. An inability to be is a modification of an ability to be that is lost. Being unable to fly, being unable to breathe under water, and so on, are not examples of being unable to be. Otherwise the concept would be too broad and we would be more unable than able to be. It is a *lost* ability or an ability that is never achieved, viewed against a background of a common ability. Being unable to be is therefore intimately linked to an ability to be, and vice versa. Being able to be is not infinite, unlimited. It is a way of existence that is granted temporarily, for a number of years, and is never assured. It is a fragile, transient gift.

Considering inability to be is one way of expressing this aspect of being able to be. Even in cases of extreme physical disability there may be a possibility of freedom of thought, imagination, emotion, and intellect. Freedom and imagination can enable even those who are unable to be in one way to be in a new way, although there are instances, such as extreme or chronic pain, in which it is not possible to enjoy one's freedom or imagination. In those cases one's ability to be is indeed radically and inconsolably curtailed.

3.5 Conclusion

This chapter explored the body in illness, examining changes to spatiality, social relations, and agency which happen in illness. It suggested that illness is characterized primarily by the losses it inflicts on the ill person. It provided some examples of the ways in which losses are experienced, and later incorporated into a modified embodiment. The chapter then examined how spatiality and social relations change in illness. Illness often brings about restrictive constraints that curtail bodily freedom and social spontaneity. The chapter suggests that Heidegger's notion of existence as 'ability to be' ought to be reframed in light of less-than-ideal ways of being such as those experienced in illness and ageing.

Acknowledging an inability and learning to see it as part of life's terrain are important lessons that illness can teach. This knowledge enables the ill person to embrace the unable self as part and parcel of human existence. By having a more balanced view of life and its challenges as interplay of ability and inability, illness can become more accepted and less disruptive.

This new understanding of human life as being both able and unable to be, paves the way to understanding three important aspects of illness: first, the question of acceptance and of living (well) with illness, which is explored in Chapter 6. Second, the notion of 'inability to be' in its extreme implies the closing down of all possibilities, namely, death, which marks the horizon of illness and all life. Death is explored in Chapter 7. Third, the experience of inability to be has a particular phenomenology, which I characterize as 'bodily doubt'. I now turn to examine this feature of illness.

4

Bodily Doubt

The deepest human problems lurk behind the obvious.

W. Blankenburg and A. L. Mishara, First Steps
toward a Psychopathology of 'Common Sense'

In this chapter I examine a core experience of illness: the experience of bodily doubt.[1] The experience of doubt is common. We often doubt facts about the world, but we also doubt our abilities, faculties such as understanding and judgement, and other psychological features, such as our strength of character, determination, and self-control. However, the experience of doubt in such cases is mostly localized, not pervasive. We might doubt our ability to run a five-kilometre race, but not our ability to run *tout court*. We might worry about a decision we have made, but not doubt our ability to make decisions generally. Indeed, we consider pervasive, persistent, and severe doubt to be a psychiatric symptom meriting medical attention; for example, the doubts experienced by people suffering from obsessive compulsive disorder (OCD).

The experience of doubt takes on a particular inflection in cases of mental and somatic disorder. For example, some people with depression seem to describe a sense of inability in an intensified and more pervasive form:

When I'm depressed, every job seems bigger and harder. Every setback strikes me not as something easy to work around or get over but as a huge obstacle. Events appear more chaotic and beyond my control: if I fail to achieve some goal, it will seem that achieving it is forever beyond my abilities, which I perceive to be far more meagre than I did when I was not depressed (Law 2009, p. 355).

---

[1] This chapter is reprinted with kind permission from the *Journal of Consciousness Studies*. It was first published in 2013 in the *Journal of Consciousness Studies* 20(7–8): 178–97.

Here is another example, this time of a somatic illness. If you have ever been very ill with the flu you may have been lying in bed and thinking about something you need to do, maybe grocery shopping. You may contemplate executing the task, but at the same time feel deep doubt about whether you could actually perform the task. A sense of doubt about a routine activity pervades you. Not only does the sense of doubt accompany the conception and planning of a routine activity, it can also colour the experience itself: walking to the shop, wandering the aisles, and so on. Unlike some forms of doubt, the sense of bodily doubt does not quickly and quietly disappear once the activity has begun. Rather, it remains constantly there, weighing down every step.

In this chapter I develop a phenomenology of this type of feeling, focusing on a particular feeling which I characterize as bodily doubt. I suggest that we have an ongoing tacit certainty about our bodies, e.g. that we will be able to digest our lunch, that our hearts will carry on beating, that our legs will carry us, etc., and that this certainty is not rationally justified. Similar to Hume's critique of our instinctive faith in induction, I suggest that a sense of unjustified and nonetheless powerful faith underlies our relationship with our bodies.

This feeling of certainty we have about our bodies can be construed as a belief we cannot help but hold, but are unable to epistemically justify in any strong sense. Indeed, this sense of bodily trust or confidence is so tacit that, at least for young healthy persons, its very existence goes unrecognized. This is a point that is fundamental to the phenomeno-logical method proposed here: we are typically oblivious to aspects of our embodied existence. The aim of phenomenology is to bring these to light. In this chapter I attempt to do just this for the tacit sense of bodily trust and for its breakdown in illness, which I call bodily doubt.

I suggest that the case of illness, in which this certainty breaks down, can illuminate the normal feeling of certainty, which is tacit and thus difficult to scrutinize.[2] Illness reveals that the kind of belief that underlies our relationship with our bodies is a bodily feeling anchored in our animal nature. I characterize bodily doubt as a radical modification of our bodily and other experience. I analyse this modification as made

---

[2] Just a reminder that I use the term illness to denote serious, chronic, or life-threatening illness, rather than common, transient illness such as the common cold.

up of three changes to experience: loss of continuity, loss of transparency, and loss of faith in one's embodied existence.

The structure of the chapter is as follows. In the first section I present bodily certainty and discuss its main features: it is unjustified and tacit. In the second section I propose that cases of breakdown of bodily certainty are characterized by a special kind of doubt, namely, bodily doubt. I conclude by presenting the philosophical importance of studying cases of pathology (mental and somatic disorder, as well as other kinds of experience), suggesting that illness is a philosophical tool for the study of otherwise hidden aspects of human experience. This final theme is one I come back to in the final chapter of the book.

4.1   Bodily Certainty

In *Feelings of Being* Matthew Ratcliffe (2008a) describes existential feelings, which are the background feelings underlying our existence. Existential feelings are a pre-given experiential context in which particular intentional attitudes are possible. On Ratcliffe's account, existential feelings ground the sense of belonging to a world, a sense of reality, which characterizes normal (2008a, p. 41). In his account Ratcliffe relies on Heidegger's mood analysis to provide an account of a feeling that grounds human existence and gives it a sense of stability and normality (2008a, pp. 47–52). The common state of affairs is one in which the world appears familiar, its workings are mostly predictable, and one feels grounded in it. This does not mean that an individual feels happy or satisfied (2008a, p. 41). Rather, the focus is on the tacit sense of reality (as opposed to unreality) and understanding of one's being in the world.

Importantly, because this feeling is usually tacit (we do not normally go about thinking how lucky we are that our perceptions are orderly and familiar; that our bodies obey us; that the world continues to appear to us, and so on), it mostly goes unnoticed and is not the object of explicit scrutiny. Husserl similarly presents the world as the 'universal ground of belief pregiven for every experience of individual objects' (1997, p. 28). He writes in *Experience and Judgement*: 'an actual *world* always precedes cognitive activity as the universal ground, and this means first of all a ground of universal passive belief in being which is presupposed by every particular cognitive operation' (1997, p. 30).

Existential feelings are distinct from more familiar types of emotion and mood, since to feel particular emotions or be in a particular mood one must already find oneself in a world—a meaningful structured whole of actions, persons, and things—against which purposive activities and our intentional responses to them become possible. Because they are often tacit, existential feelings are difficult to study and not often the focus of a philosophical analysis.

We lack a sophisticated vocabulary for existential feelings and they are often conflated with other, distinct types of mood and emotion. Indeed, Ratcliffe's monograph is the first of its kind. He claims that existential feelings can best be studied by looking at pathological cases in which the feeling has been disturbed. Ratcliffe examines such cases, for example the Capgras delusion (a rare psychiatric disorder in which a person thinks that friends and family members are impostors) and Cotard syndrome (another rare disorder in which a person thinks they are dead or that their bodies are rotting, or that their internal organs have been removed). Ratcliffe also discusses more common disorders such as depression and schizophrenia, in which the sense of reality is disturbed.

Ratcliffe claims that these feelings of being are *bodily*, but also give a sense of how the world is and of one's relationship to the world. He writes: 'bodily feelings are not just feelings of internal bodily states; they can also contribute to experiences of things outside the body'. 'Certain feelings [. . .] are ways of finding ourselves in the world, existential backgrounds that shape all our experiences' (2008a, p. 47). This dual role of existential feelings is of importance to Ratcliffe, as it synthesizes the internal and the external, feeling and cognition.

A particular aspect of existential feelings I would like to focus on here is the feeling of bodily certainty or of bodily doubt that accompany existential feelings. I suggest that a feeling of bodily certainty or uncertainty is a necessary constituent of existential feelings that makes its bearer present in a world by offering her a meaningful horizon in which things and projects can appear. In other words, different degrees and kinds of bodily certainty and uncertainty are integral to all existential feelings.

Let us look at bodily certainty first. We feel tacitly confident (or rather, we do not normally question) that our bodies will continue to function in a similar fashion to the way in which they have in the past. We expect our stomachs to digest the lunch we have eaten, our brains to continue to process information, our eyes to continue to see, and so on. This feeling is normally so tacit that it is difficult to describe. Revealing it is the task of this section.

By bodily certainty I denote the subtle feeling of 'I can' that pervades our actions (compare Husserl 1988). This is the feeling of possibility, openness, and ability that characterizes routine and familiar actions. This feeling is anchored in what Merleau-Ponty (2012) called the 'habitual body', the body's accumulated habits and routines. The 'I can' feeling implicitly underlies all our movements and plans. If I plan to walk the dog, implicit in that plan is my confidence that I am able to walk. This 'I can' feeling is not unlimited. We do not have such a feeling when we contemplate flying or breathing under water (see the final section of Chapter 3). But this feeling of possibility and freedom does characterize our attitude towards most of our actions. Even if I plan to run a half-marathon and doubt whether I would be able to complete the course, this doubt is more of a challenge or motivator than the negation of the 'I can' feeling. In other words, such a doubt does not disrupt the feeling of bodily certainty, but rather extends the sense of certainty by challenging us to develop further abilities. Indeed, challenging bodily activities such as competitive sports define themselves as explicitly pushing the limits of bodily performance and its associated psychology.

Normally bodily certainty is tacit, but it can be made explicit and become the object of philosophical reflection. When I type these words I am oblivious to the speed and expertise with which my fingers find the needed keys. But I am easily able to turn my attention to this achievement and pay explicit attention to it (although this disturbs the fluidity of the typing, which lends support to the idea that the knowledge involved is tacit).

Most of the time when I type, I pay no attention to my typing and take for granted the speed, ease, and painlessness of this activity. I know very little about the neurology and physiology involved in typing and have little interest in typing over and above its practical use to me. This kind of taken-for-grantedness—unreflective, disinterested—characterizes bodily certainty.

When my husband lifts our son onto his shoulders and runs down the path, he does not consider this an achievement. It is just something he does in the full, but tacit, confidence that he can bear the child's weight, that he is able to run, keep his balance, and so on. He has been able to do this previously, so takes for granted that he will be able to do this now. This type of bodily certainty is pervasive and provides the background certainty that underpins our attitudes towards familiar actions.

This way of talking about bodily certainty may be misleading. It makes it seem as if this feeling is constantly monitored and reflected upon; as if it is an explicit part of experience. But actually most routine actions we perform—walking, talking, eating, and so on—are rarely consciously reflected upon, rarely thematized, let alone problematized. Rather, they form a transparent background that is just there, in stable form, often for many decades.

The sense of certainty, therefore, is not conscious but pre-reflective. It is the immediacy and automatic manner in which we turn to familiar tasks. It is a basic mode of action that takes agency for granted. This certainty is most effective when it is unnoticed. In the same way that we do not pause to wonder whether our food was poisoned by the chef in a restaurant (compare Bernstein 2011), we do not need to check that we know how to chew. Indeed, this would be a cumbersome and unnecessary procedure much of the time. We simply have a sense of certainty and usually also of bodily ease and familiarity that go unnoticed and yet underpin action.

This feeling has a phenomenology that can be described when we turn our attention to it, as well as propositional content, although bodily certainty is not primarily a propositional attitude. It is only when we explicitly reflect upon this certainty, or express it to others, that a propositional attitude is involved. This propositional attitude is secondary to a more fundamental, non-propositional form of experience. The phenomenology is straightforward: we feel able, confident, and familiar with what we are doing. We not concerned with the possibility of bodily failure, and we simply perform the task at hand. Moreover, this is primarily a *bodily* feeling. The tacit aspect of the feeling lies here: our body proceeds with pre-reflective confidence and habitual ease; there is no need for reflection.

The sense of certainty is primarily to be found in the tacit confidence with which we make a cup of tea, slice an apple, walk the dog, or dial a number. There is no reflection, and there is a clear sense (again, tacit) that the action is meaningful, familiar, and possible. This taken-for-granted confidence also makes the propositional content seem artificial and uninteresting. Normally we are not called upon to formulate the feeling of certainty or pay it attention. It is simply there, as a silent enabler of action.

Bodily certainty serves as the ground on which we base many of our assumptions and expectations. What an analysis of it reveals is that our most abstract goals and assumptions are based on a bodily feeling of ability giving rise to an existential sense of possibility. This certainty, lack of need to attend to our bodies, is core to our being. This fundamental role of bodily certainty can be seen as analogous to the role of hinge propositions (Wittgenstein 1974, §341). These propositions are characterized by the fact that they are exempt from doubt and are like hinges on which questions and doubts turn (1974, §341). The 'hinge' in the case of bodily certainty is not a proposition, but a bodily feeling that can be expressed as a proposition, but is not primarily cognitive, of being in control of one's body, having a sense of familiarity and continuity with respect to one's body. A tacit existential feeling of trust (compare Bernstein 2011), familiarity (compare Ratcliffe 2008a), and normalcy underlies our everyday activities and actions.

4.2 Bodily Doubt

One way to make visible the bodily certainty that underpins ordinary human experience is by looking at cases of its breakdown, namely, illness. In many somatic and mental disorders the sense of certainty and confidence we have in our own bodies is deeply disturbed. Basic tacit beliefs about bodily abilities that were previously taken for granted are suddenly, and sometimes acutely, made explicit and thrown into question. Such cases of illness make apparent not only the bodily feeling of confidence, familiarity, and continuity that is disturbed, but also a host of assumptions that hang on it. For example, one's future plans depend on bodily capacities and thus are limited by ill health. The ill person's temporal sense is radically changed by a poor prognosis. Her values and sense of what is important in life may be profoundly modified in light of illness; bodily limitations impact on existence generally (Carel 2013a). The ill person's concepts undergo a radical change; for example, the meaning of terms such as 'near' or 'easy' may fundamentally change (Carel 2012).

Bodily doubt is not just a disruption of belief, but a disturbance on a bodily level. It is a disruption of one's most fundamental sense of being in the world. Bodily doubt gives rise to an experience of unreality, estrangement, and detachment. From a feeling of inhabiting a familiar world, the ill person is thrown into uncertainty and anxiety. Her attention is withdrawn from the world and focused on her body. She may feel isolated from others, who maintain their connection to the world, and may become detached from both physical and social environments. The natural confidence in her bodily abilities is displaced by a feeling of helplessness, alarm, and distrust in her body.

There are different degrees of bodily doubt. It may vary in duration, intensity, and specificity. Part of our experience of the flu, for example, is the understanding that it is temporary. Thus an experience of bodily doubt associated with the flu differs from bodily doubt experienced in the context of Parkinson's disease, for example, or another serious chronic or progressive condition. Bodily certainty can be regained in some cases, but the expectation that the doubt may return substantively changes the kind of experience involved in each case.

Similarly for intensity: a bad case of tonsillitis may involve bodily doubt at the time without a longer-term shift in the structure of one's experience.[3] A more radical doubt, where the intensity of the experience is greater and the feeling is expected to return, may profoundly change the structure of one's experience.

Finally, bodily doubt can be all-pervasive, or it may relate to specific aspects of bodily functioning. It is possible to experience doubt about a certain action (walking upstairs) without calling into question other, even interrelated, aspects of bodily existence (such as balancing). I suggest that this wide range of experiences share common features that make the experience of bodily doubt philosophically revealing by exposing the structure of how things normally are.

Bodily doubt has additional features that make it phenomenologically significant and philosophically revealing. First, bodily doubt can descend at any moment. In some cases its appearance may be gradual, but it may come upon a person with complete surprise (injury-causing accidents are such cases). The body's precariousness and unpredictability make it more threatening and us less capable of incorporating our experience of it into our familiar world.

Second, the feeling of bodily doubt invades the normal sense of things. It results in feeling exposed, threatened. It is uncanny and may give rise to a kind of anxiety. In this respect bodily doubt is different to normal

---

[3] Less serious conditions can also be philosophically revealing, as discussed in Chapter 2, §2.4.

kinds of bodily failure (tripping up, failing to learn a dance move, feeling too exhausted to go to a party) that may seem similar to bodily doubt.

As was argued in Chapter 2, §2.4, when experiencing normal failure one is still immersed in the world. The failure is incorporated into one's current existential feeling, which is often an undifferentiated background feeling. In contrast, bodily doubt disrupts the normal sense of being in the world and replaces immersion with suspension. In normal conditions one may experience bodily failure that may be frustrating and humiliating (if you have ever failed to fix a bicycle puncture, you will know the feeling) but one remains immersed in the task and in the world. Bodily doubt casts us out of immersion and into suspension; the familiar world is replaced by an uncanny one.

Is bodily doubt a type of anxiety and hence a type of mental disorder? What picks out experiences of bodily doubt from other types of anxiety is that they are experienced as *embodied* doubt about bodily capacities normally taken for granted. Heidegger described anxiety as a state of loss of meaning (1962). But bodily doubt, in contrast, has meaning: the meaning is the doubt itself. The doubt is neither irrational nor meaningless. It can be clearly explained and expressed.

Bodily doubt can be helpfully contrasted with panic attacks, which have a similar phenomenology. While the experience of bodily doubt and of a panic attack may be similar (acute sense of disruption, anxiety about one's ability to breathe or stay conscious, etc.) the two can be distinguished on non-phenomenological grounds. Bodily doubt is situationally appropriate and associated with true beliefs, whereas panic attacks are situationally inappropriate and frequently associated with false beliefs (e.g. a young, healthy individual thinking, 'I am about to die').

Third, bodily doubt reveals the extent of our vulnerability, which is normally masked. Once experienced, bodily doubt often leaves a permanent mark on the person experiencing it; it is the loss of a certainty that has hitherto not been disturbed. Moreover, the *possibility* of it being disturbed has not been countenanced. In this sense it is like other kinds of trauma (being the victim of violence, bereavement, being involved in an accident, and so on). What these cases have in common with bodily doubt is that they overthrow our most basic assumptions about the regularity, predictability, and benevolence of the world.

Jay Bernstein describes trauma victims as undergoing an experience 'revealing underlying and intractable dimensions of vulnerability, dependence, and potential helplessness that are normally hidden from consciousness' (2011, p. 399). Bodily doubt similarly reveals vulnerability of a specifically bodily kind. Illness can also be likened to a loss of innocence; it is impossble to return to the naïve (and, in retrospect, gullible), state of confidence one was in before. There is no turning back once genuine bodily doubt has been experienced; one's basic orientation in the world has changed and the possibility of catastrophic bodily failure is now part of one's experiential horizons.

Fourth, bodily doubt makes the person experiencing it feel incapable. Confronting the loss of abilities and the frustration involved in 'being unable' ('I cannot') contrasts with the normal (healthy) feeling of competence and ability, even when this ability is punctuated by occasional failure. For example, I decide to go for a walk after dinner to stretch my legs. Of course I cannot walk indefinitely, but my bodily certainty is such that the project (going for a walk) dominates the action. I walk until I am satisfied, so in effect I have walked without limit. In a state of doubt I want to go for a walk, but I am restricted by my bodily limitation. I plan before I act (how far I can go; will it be too steep? what if I get too tired?), such that the action becomes determined by my bodily limits. The limitless sense of myself, of my open horizons, as extending beyond myself and into the world collapses back onto my actual physical being; it becomes an act of conscious planning to see how far I can, or dare, project myself onto the world.

An important question is what is the relationship between bodily doubt and bodily incapacity? Bodily doubt can be experienced in the absence of a known bodily incapacity, as might be the case in depression or panic attacks. Bodily doubt can also diminish as one adjusts to certain incapacities (e.g. in certain forms of disability) and may be completely absent in the case of congenital disability. So there are grounds for differentiating between states of 'being unable' and states of not knowing, or suddenly feeling doubtful about whether one might be able to. I suggest that bodily doubt is part of serious illness and constitutes the transition from health (bodily capacity) to illness (bodily incapacity), via the experience of bodily doubt. In anxiety disorders, where there is no underlying somatic illness, the disorder can be described as a series of non-justified experiences of bodily doubt that repeatedly reoccur because there is no transition to bodily incapacity.

More generally, the sense of inability may stem from a bodily dys-function (e.g. loss of mobility) but also from an incapacitating mental disorder, such as depression.[4] Depression often gives rise to severe feelings of helplessness and ineptitude, which can be crippling. More-over, the physical symptoms of depression can be severe and comparable with somatic inflammatory disease (Ratcliffe et al. 2013). Bodily doubt affects the spectrum of possibilities available to a person in the practical sense (being unable to perform an action) as well as in the existential sense (perceived narrowing down of possibilities). But it does not simply narrow down possibilities *tout court*. Some types of possibility are lost at a faster rate, or to a greater extent in illness.

It is not just a narrowing down of possibilities *simpliciter*, but certain types of possibilities, contingent upon the nature of the illness, become entirely dead, while others might be maintained in modified form. Bodily doubt challenges the everyday veneer of normalcy and control that we cultivate as individuals and as a society. As such it reveals the true state of affairs, which is one of vulnerability, existential uncertainty, and depend-ency on others (compare Bernstein 2011).

Phenomenologically speaking, bodily doubt is experienced as anxiety on a physical level, hesitation with respect to movement and action, and a deep disturbance of existential feeling. The doubt distorts the sense of distance and time, in a way akin to other kinds of psychological stress. However, the important feature of bodily doubt is that it is not a mere psychological state, although it certainly has psychological features. It cannot be reduced to a mental state or propositional attitude. Even in the case of bodily doubt arising from a mental disorder such as depression, it is still a bodily feeling.

Bodily doubt is a physical sensation of doubt and hesitation arising in one's body. It is not solely cognitive, although it can be expressed in propositions. Here is an example. Patients with severe respiratory disease commonly report fearing that they will be unable to breathe. This feeling cannot be described satisfactorily as a panic attack because in these cases there is rational ground for the fear. The fear stems from a physical experience of acute breathlessness that at its extreme leads to respiratory failure. In such cases bodily doubt is rooted in a disease so severe that it

---

[4] Bodily doubt could also precipitate mental disorder, as well as be triggered by it. I thank Ian James Kidd for bringing this to my attention.

throws into doubt the belief that one will continue to breathe. This is, of course, terrifying; but the psychological features of this experience are a result of the bodily doubt, not the doubt itself.

Bodily doubt not only changes the content of experience: it also pierces the normal sense of bodily control, continuity, and transparency in a way that reveals their contingency. It shows our tacit faith in our own bodies to be a complex structure that becomes visible when it is disturbed. It destroys the normal experience of continuity, transparency, and trust that characterizes this structure, which I now discuss in more detail.

1. Loss of continuity: human cognition and action are characterized by continuity of experience and purposeful action (Sartre 2003; Merleau-Ponty 2012; Noë 2004). In bodily doubt this continuity is suspended and replaced by a modified awareness of self and environment. In this suspension experience becomes discontinuous. The characteristic smoothness of everyday routine is disrupted. Everyday habits become the object of explicit attention and conscious effort; the ongoing tacit sense of normalcy is lost. In this situation one is unable to pursue their goals because the normal flow of actions leading to the goals is disturbed. Minor tasks require planning and attention to detail, as well as contingency plans. The normal flow of everyday activities is halted by bodily uncertainty and when it is resumed it is altered by the experience of doubt. The possibility of doubt taints further experience, even if continuity is restored. The possibility of doubt is a constant reminder of the contingency and fallibility of the original continuity.

In mental disorder, in particular in anxiety and depression, the pattern is different. These conditions may lead to a state characterized by Heidegger (1962) as *Angst*, or anxiety. In anxiety one's sense of purposeful activity is lost, leaving the person unable to act. Action is grounded in meaning: I tap keys on my computer keyboard in order to write this book. I write this book in order to convey certain ideas. I convey certain ideas in order to contribute to a debate in philosophy, and so on. But, as Heidegger points out, ultimately, this nested set of goal-directed activities comes to an end.

Ultimately human existence is ungrounded. This realization leads to what Heidegger calls *Angst*. In *Angst* purposefulness is removed and with it the meaning of entities. They turn from being ready-to-hand (*Zuhanden*) entities we use (keyboard, desk, reading lamp) to being present-at-hand (*Vorhanden*) entities which confront us with their lack of usefulness, and hence their lack of meaning. In anxiety the intelligibility of the world surrounding us is lost, because the practical coherence of entities has been lost once the sense of purposefulness is gone.

This experience has been documented by people with mental illness. Ratcliffe (2013) discusses the sense of unreality that characterizes schizophrenia, citing a patient, who says:

> When, for example, I looked at a chair or a jug, I thought not of their use or function—a jug not as something to hold water and milk, a chair not as something to sit in—but as having lost their names, their functions and meanings (cited in Ratcliffe 2013, p. 85).

The loss of meaning that accompanies experiences of *Angst* arises because a person is suddenly severed from the inherited everyday structures of meaning and value that the 'They' (*das Man*) inhabit—of what *They* do and what *They* value, etc.—and so everything 'sinks away', as one is no longer 'absorbed' and 'everyday familiarity collapses' (Heidegger 1962, pp. 232ff).

The role of *das Man* is important to Heidegger's analysis of *Angst*, but it is also different, to some degree at least, in the case of somatic illness. For in illness a person might still be 'absorbed' in the inherited, unreflectively inhabited structures of meaning and value of *das Man*; they may be completely absorbed in applying for a job, or raising children, or another project. It's just that suddenly the ill person's capacity to pursue these projects is radically disrupted; she might still want those things but find them no longer within reach.

If in such cases there is not the actual 'sinking away' from 'everyday familiarity' of coping absorption, then we must ask whether somatic illness is also a full experience of *Angst* in Heidegger's sense. I suggest that such a radical break experienced within full absorption in a project necessarily withdraws the ill person from everyday familiarity and forces her to extract herself and view her position as no longer compatible with that of *das Man*. In the way that language is no longer shared (terms such as 'near' and 'easy' change their extension in illness), also shared understandings about preferences, utility, and agreed pragmatic uses of items in our environment lose their sharedness, and hence lead to a loss of meaning. Bodily doubt *necessarily* prompts the ill person to critically renegotiate her participation in the shared world of public norms. And it thus seems that Heidegger's account of *Angst* does apply in somatic illness, as well as in mental disorder.

The loss of continuity also leads to a break with past abilities, which are not guaranteed to return in the future. Indeed, reliance on past abilities is broken by bodily doubt. This brings about a loss of habits and expectations which make up one's everyday dealings with the world and requires broad adjustment to the new conditions (Carel 2013a). The outcome of such adjustment may be reducing one's engagement with the world, social withdrawal, and overall reluctance to go beyond the narrow band of activities considered safe.

Finally, whereas bodily certainty synthesizes present and future through goal-directed action, in doubt the present moment (incapacity) is split off from the future (goals). In doing so, living bodies that are future-directed, world-facing and projected become fragile physical entities, inert, self-facing, that are non-projecting and dislocated from embodied action. The body turns from being a ready-to-hand entity, poised to act and immersed in a world, to a present-at-hand uncanny entity that has lost its capacity for intentional action and is suspended from the world. The focus shifts from experiencing oneself primarily as an intentional subject to experiencing oneself as a material object.

2. Loss of transparency: As we have seen, Leder and others characterize the healthy body as transparent (Sartre 2003) or even absent (Leder 1990). Leder writes: 'while in one sense the body is the most abiding and inescapable presence in our lives, it is also essentially characterized by absence. That is, one's own body is rarely the thematic object of experience' (1990, p. 1). As discussed in Chapter 2, this transparency is somewhat idealized in philosophical descriptions of health, since even this transparency is often pierced by experiences in which the body comes to the fore, sometimes in negative ways. Illness, in contrast, creates areas of dramatic resistance in the exchange between body and environment. In cases of bodily doubt, the body's taken-for-granted capacities become explicit achievements. What was previously per-formed with little or no thought now requires conscious planning. For example, in the case of bodily doubt, shopping for groceries may become a full-blown project. Moreover, the action is understood in terms of its limits which also leads to a loss of spontaneity and changes the meaning of routine tasks.

Additionally, the body becomes explicitly thematized as a problem. The tacit taken-for-granted attitude we have towards it (we expect our bodies to perform complex actions, to be pain-free, to allow us to concentrate, and so on) is replaced by an explicit attitude of concern, anxiety, and fear. In bodily doubt we may even worry about aspects of our body that are normally invisible (e.g. liver function) and alter our behaviour accordingly. Merely knowing about a particular risk associated with an illness, whether real or imagined, is often enough to modify bodily habits so they are slower, more hesitant, or otherwise censored. This, too, is a kind of bodily doubt.

The psychological consequence is harmful: the attitude of the ill person towards her body is negative and this appraisal is often reinforced in the medical encounter (Carel 2012). For example, most test results measure how much certain organ functions deviate from the normal range (e.g. kidney and lung function are measured as percentage of predicted value). Medical encounters usually focus on the dysfunction at hand, thus becoming unpleasant reminders of bodily incapacity or disease progression. These encounters contribute to the explicit thematization of the body as a problem and reduce the ill person's ability to experience the body as transparent.[5]

Finally, there is a melancholic element to this loss of transparency. The body as it was before is often viewed nostalgically, as a lost era (ageing is a similar case) and the experience of adjusting to this loss has been likened to mourning (Little et al. 1998). The transparency and spontaneity of bodily certainty is replaced with opacity and passivity (Sartre 2003, p. 359). To paraphrase Sartre, bodily doubt 'fastens on to consciousness with all its teeth', and the body becomes a source of uncertainty, pain, and suffering (p. 360).

3. Loss of faith in one's body: The loss of faith pertains to the tacit set of beliefs we hold about our bodies. These beliefs support everyday actions as well as more specialized goals and projects. Whatever the action, it is hard to carry it out in the context of doubt. Not being sure one is able to achieve the simplest of tasks leaves one in turmoil and damages the

---

[5] I am not critical of the medical need to measure dysfunction, but I think it is important to appreciate the effect this has on the ill person.

implicit certainties underpinning the everyday.[6] This loss of faith is not an experience of mental breakdown, although it could be triggered by some kinds of mental disorder. The phenomenology of these experiences is so firmly rooted in the body that it would be misleading to say that they are some kind of mental error. The loss of faith is a way of experiencing one's body which replaces the lost certainty. This is an experience of vulnerability and hesitation experienced on a bodily level. This experience amounts to a disruption of one's sense of belonging to the world and the disappearance of the sense of ordinariness.

The loss of faith in one's body reveals the contingency and fallibility of our normal trust in our bodies. In this sense, the loss of faith as the failure of bodily certainty makes explicit the weak epistemic status of our everyday beliefs. I suggest that bodily doubt is a bodily enactment of Hume's critique of induction. Bodily doubt reveals the bodily certainty previously taken for granted and inductively learned to be epistemically unjustified. Hume writes:

As to past *Experience*, it can be allowed to give *direct* and *certain* information of those precise objects only, and that precise period of time, which fell under its cognizance: but why this experience should be extended to future times, and to other objects, which for aught we know, may be only in appearance similar; this is the main question on which I would insist (1975, pp. 33–4).

Hume concludes that 'wherever the repetition of any particular act or operation produces a propensity to renew the same act or operation, without being impelled by any reasoning or process of the understanding, we always say, that this propensity is the effect of *Custom*' (Hume 1975, p. 43). I suggest that such custom underlies our tacit beliefs about our body's continued functioning and our sense of bodily certainty. Hume comes close to stating this when he asks: 'The bread, which I formerly eat, nourished me [. . .] but does it follow, that other bread must also nourish me at another time [. . .]' (p. 34). Finally, in the *Dialogues Concerning Natural Religion* Hume also argues (through the words of Philo) that 'vulgar' prejudices, which are continually reinforced

---

[6] 'Turmoil' (*tarachê*, which literally means trouble, disorder, or confusion) is the term ancient sceptics used to characterize discrepancies in their impressions that lead to doubt. For the sceptic, investigation leads to suspension of judgement, which resolves the turmoil and brings about *ataraxia* (<http://plato.stanford.edu/entries/skepticism-ancient/> accessed 11 August 2014). I thank Giles Pearson for clarifying the Greek term for me.

by experience, must be respected 'since [such prejudice] is founded on vulgar experience, the only guide which you profess to follow' (Hume 2008, Part Six). Bodily certainty could be seen as such a 'vulgar prejudice'.[7]

It is interesting to note that Hume wrote, in a famous passage of section 7 of his *Treatise of Human Nature*, that 'philosophic doubt' was deeply embodied, and best dispelled not by reasoned disputation, but through bodily and social activity:

Most fortunately it happens, that since reason is incapable of dispelling these clouds, nature herself suffices to that purpose, and cures me of this philosophical melancholy and delirium, either by relaxing this bent of mind, or by some avocation, and lively impression of my senses, which obliterate all these chimeras. I dine, I play a game of backgammon, I converse, and am merry with my friends; and when after three or four hours' amusement, I would return to these speculations, they appear so cold, and strained, and ridiculous, that I cannot find in my heart to enter into them any farther (Hume 1896).

I suggest that bodily certainty is akin to the epistemically ungrounded tacit beliefs Hume argues we hold. The loss of faith experienced in bodily doubt reveals the tenuousness of this belief. The tacit belief in the continued functioning of our body enables us to pursue everyday goals and plans. We hold other kinds of tacit beliefs, e.g. that our house is still standing although we are now in the office. The question whether these beliefs are irrational, or how powerful Hume's critique is has been discussed at length (Okasha 2003). The analogy to Hume I wish to draw here is to the *fallibility* of such beliefs, which contrasts sharply with their powerful hold on our minds and behaviour.

As Bernstein (2011) points out, there is a fundamental sense of trust which underlies our commerce with the world and this tacit network of trust relationships enables us to go about our business without being paralyzed by doubt. On Bernstein's view, trust is a non-propositional attitude, a basic feeling that permeates our interactions with other people. What this trust enables us to do is to find ourselves in a world that is peaceful and stable enough to enable us—amongst other things— to philosophize. Husserl comments in a similar vein: 'it is this *universal ground of belief in a world* which all praxis presupposes, not only the praxis of life but also the theoretical praxis of cognition' (1997, p. 30).

---

[7] I thank Andrew Pyle for pointing me to this source.

Lived doubt, of which bodily doubt is one instance, is pervasive and disrupts everyday action. Thorsrud (2009) claims that 'all intentional, purposeful action presupposes some sort of belief' (p. 173).[8] If action depends, in part, on belief, then the suspension or loss of belief must disrupt action. As such bodily doubt goes to the heart of everyday life.

4.3 Pathology as Method

So far, this chapter has explored the systematic invisibility of bodily certainty. To paraphrase Bernstein (2011) on trust, bodily certainty best realizes itself through its disappearance. In order to study this certainty we should turn to cases in which it is absent, namely, cases of mental or somatic disorder. What can pathologies tell us about the normal way of things? If normally I am tacitly certain that my body will continue to support me, what does a disruption of this cause?

Bodily certainty is part of a tacit sense of belonging to the world that is normally unnoticed (Ratcliffe 2008a). What cases of pathology enable us to see is that what ordinarily binds you to the world can become visible and hence the object of study. I would like to end by suggesting that certain types of pathology are crucial for the philosophical study of human existence (a theme further developed in Chapter 9). In such cases the body becomes unnatural and is thematized in new ways. This enables us to explore dimensions of it that would normally go unnoticed.

The term 'pathology' may include other kinds of alienated or explicitly thematized embodiment. For example, in her essay 'Throwing like a girl' Iris Marion Young (2005a) describes embodiment that is experienced as unnatural and gazed upon. Her study of the cultural pathologization of female embodied experience includes phenomenological analyses of aspects of embodiment that are both fundamental and mundane (e.g. the pregnant body and the portrayal of women in advertising) (Young 2005b).

Pathological cases of embodiment provide a philosophical tool with which to illuminate normal embodiment. But is this tool available to everyone, or only to those who experience pathology? Is the first-person experience of bodily doubt a prerequisite for certain kinds of philosophical

---

[8] But note that 'belief' is too often treated as a unitive concept—*to believe that* p—in a way that disguises the epistemological and phenomenological diversity and complexity of the range of things collectively described as 'believing'.

insight? Or should we maintain that second- or third-person access is sufficient? If one lacks the first-person experience, how does one come to appreciate the relevant phenomenology? I suggest that in order to glean philosophical insight from bodily doubt one does not need to have the first-person experience herself.

There is good evidence that empathy can train and redirect modes of thinking, moral insight, and sensibilities in philosophically important ways (e.g. Nussbaum 1990; Ratcliffe 2012b), so the philosophical method proposed here is universally available. The philosophically relevant features of bodily doubt can be conveyed by using literature, film, testimonies, and so on. However, it may be true that there is something ineffable about radical bodily experiences that can only be fully appreciated if one undergoes the experience oneself (childbirth is a common example used to support this claim, e.g. Heyes 2012). It may be that some of the experience cannot be propositionally expressed. This is a limitation that is inherent to this method, but does not seriously hinder it. The method is still widely accessible and useful across a range of philosophical issues.

Returning to Merleau-Ponty's (2012) study of Schneider (based on Gelb and Goldstein's case study), we can see this method at work. Schneider suffered a brain injury which restricted his actions in unusual ways. He lived a seemingly normal life, working and living independently. However, he was limited in an interesting and pervasive way. He was unable to change his daily routine (e.g. stray from his normal route home). He could not initiate sexual relations although his sexual function was intact. Merleau-Ponty characterizes Schneider's malaise as existential, arguing that Schneider's ability to initiate, imagine, and fantasize, has been lost (2012, p. 135).

For Merleau-Ponty, Schneider's peculiar pathology[9] reveals our bodies as 'the potentiality of a certain world' (2012, p. 106). What we are normally able to do is to 'reckon with the possible', which thus acquires a certain actuality. In Schneider's case the field of actuality is limited and has to be made explicit through conscious effort. It is this conscious, explicit effort and the correlating achievement of action that make pathological cases useful (Gallagher 2005). By making explicit and artificial

[9] It was thought that Schneider's pathology was vision agnosia, but recently Marotta and Behrmann (2004) have argued that he suffered from integrative agnosia.

what normally goes unnoticed, such cases draw our attention to how things are normally. Pathological cases show that an experience of smooth interwoven experience and action is not a taken-for-granted ground, but an achievement (Ratcliffe 2013). Because the end product seems so natural we need pathological cases which with to expose the underlying process that gives us a normal world.

4.4 Conclusion

This chapter presented bodily certainty and its absence, bodily doubt, as a core feature of existential feelings. I described bodily doubt as a loss of continuity, loss of transparency of the body, and loss of faith in one's body. I suggested that bodily certainty is an instance of Hume's general claim about our animal nature: we feel certain that our bodies will continue to function. This certainty is not epistemically justifiable but is impossible to relinquish. The experience of bodily doubt makes this tacit certainty explicit and enables us to study it.

I argued that the exploration of pathology is useful for highlighting tacit aspects of experience that otherwise go unnoticed. Finally, I put the specific claim about the usefulness of examining bodily doubt in broader context, suggesting that it provides a philosophical tool for studying human experience. The study of pathological cases is crucial for illuminating normal function, because it brings to the fore implicit processes and assumptions that are not normally available to us. Since bodily doubt is primarily an experience arising from mental and somatic disorder, philosophy has much to learn from the study of illness.

What I hope this chapter has shown is the extent to which we are dependent on our bodies, not only for everyday functioning but also for the broader sweep of life, including our goals, plans, and expectations. Bodily certainty grounds and enables human activity. Ultimately, this ground is beyond our control and its potential fallibility is ever-present. Human existence is characterized by a brute dependence on utter unreliability, on our feeble, transient organismic structures. We put much effort into denying how feeble our bodies and minds are. Perhaps it is time to turn to this fact and study it in order to reveal the philosophical significance of human vulnerability.

5

A Phenomenology
of Breathlessness

So far this book has developed a theoretical framework for a phenomenology of illness, while engaging with the lived reality of illness in order
to generate and refine this framework. In this chapter, I want to put this
framework to a concrete test, asking whether a phenomenological analysis can reveal invisible or difficult-to-articulate aspects of a common
and debilitating symptom, namely breathlessness. I draw on first-person
accounts of others as well as on my experiences as a respiratory patient to
apply the framework developed so far to the case of breathlessness, in
order to outline a phenomenology of breathlessness. The chapter uses
the phenomenological themes developed in earlier chapters in order to
shed light on the experience of pathological breathlessness and of being a
respiratory patient.

5.1 Why Breathlessness?

Breathlessness is a common symptom of a wide range of respiratory and
cardiac disease as well as other disorders, such as panic attacks and
muscular disorders affecting the chest muscles. As such, it affects a
significant number of patients around the world. Indeed, chronic
obstructive pulmonary disease (COPD), which is characterized by progressive and debilitating breathlessness, is predicted to become the third
leading cause of death globally by 2020 (Barnes and Kleinert 2014).

Prima facie, breathlessness seems like a simple bodily sensation. In
fact, it is multifaceted and incorporates several sources of input, such as
oxygen saturation level, respiratory effort, chest muscles' work, and
blood pH levels. In severe desaturation, feelings akin to suffocation,
dizziness, sense of loss of control, and incontinence are often present.

Overall the sensation is one of acute unpleasantness and distress, but not of pain as we normally understand it.

Recent work in the neurophysiology of breathlessness shows that the same brain pathways are activated in breathlessness as in pain, hunger, and thirst (Herigstad et al. 2011). This may lead us to explore the possibility that breathlessness bears a family resemblance to pain, but is not analogous to it. It also makes salient the need for a phenomenological analysis, which may reveal how the experience of breathlessness is similar or dissimilar to other unpleasant sensations. Overall, the 'pain roadmap' proposed by proponents of the analogy with pain, seems not to provide a satisfying account of breathlessness, because of the experiential differences between pain and breathlessness. Breathlessness is not painful but acutely distressing in other ways, as explored later. Perhaps the distress is mostly affective (e.g. feeling panic) or cognitive (e.g. thoughts about loss of control and death), but there are likely more complex interactions between the different dimensions. This requires further empirical study.

Herigstad and colleagues list at least nine areas involved in the voluntary control of breathing, including cortico-limbic structures that also serve sensations such as thirst, hunger, and pain, and the amygdala (also part of the limbic system) which deals with memory and emotions (2011, p. 813). Such neuro-imaging studies have the potential to help distinguish sensory from affective components of breathlessness and improve the understanding of how emotional and cognitive processes affect not only the perception but also the pathophysiology of breathlessness.

Breathlessness can be experienced with or without other symptoms such as oxygen desaturation, hyperventilation, dysfunction of chest muscles, and dizziness. It may be that the breathlessness experienced in an asthma attack, in which the airways are suddenly narrowed, is different to the experience of breathlessness in COPD, or heart failure. Currently the term 'breathlessness' is used in both medical and lay language to point to breathlessness arising from a vast array of medical conditions and it is likely that there is significant diversity within this group. In other words, the single term 'breathlessness' covers up an experiential diversity that needs to be phenomenologically explored. The apparently simple term masks the complexity and diversity of the experiences that we call 'breathlessness'. Here I aim to uncover some of this complexity. However, a future task for a phenomenology of breathlessness is to determine whether a different taxonomy or richer language are required, and if so, to develop these in order to underpin accounts of breathlessness in general, as well as in particular diseases. One category, 'unexplained breathlessness', is of particular note here, as it consists entirely of reported experiences.

It is unfortunate that we use the same term—breathlessness—to cover both normal breathlessness experienced on exertion, which can be enjoyable, and pathological breathlessness, which is debilitating and unpleasant. Do the two sets of experiences overlap? Do they have shared features? Do they lie on a continuum or are they phenomenologically distinct? How do people who focus on their breathing (what we might call 'aware breathers') come to experience breathlessness? What can particular breathing practices, such as meditation, singing, playing musical instruments, and sports, teach us about breathlessness? Can we compare 'aware breathers' who are healthy with those who become aware of their breathing due to a pathology? Clearly there is phenomenological work to be done in this area, both philosophically and empirically.[1]

Breathlessness is a very common symptom. According to improving and integrating respiratory services (IMPRESS), 25 per cent of attendees to emergency departments, 62 per cent of elderly people, and around 95 per cent of people with COPD report breathlessness.[2] It is a strong predictor of mortality, and very common in the end stage of many diseases (IMPRESS).

Breathlessness is a common symptom, but it is also unique. Unlike many other medical symptoms, breathlessness (and breathing) has complex and powerful psychological, cultural, and spiritual dimensions. We 'take a deep breath' to calm ourselves; we are invited to 'inhale deeply' the fresh air in the countryside; a remarkable artwork 'takes our breath away'.

Breathing is deeply and intimately connected to, and reflective of, our state of mind, feelings, mood, and sense of well-being. That is why

---

[1]   A Wellcome Trust-funded project, *Life of Breath*, which I am leading with Professor Jane Macnaughton from Durham University, examines breathlessness from a multidisciplinary perspective, trying to answer some of these questions as well as bring the clinical and cultural perspectives into dialogue. See <www.lifeofbreath.org>.

[2]   See <http://www.impressresp.com/index.php?option=com_content&view=article&id=172> accessed 13 October 2015.

breathing is central to many spiritual practices such as pranayama, meditation, and religious practices, especially in Eastern religions. When one is anxious or frightened, breathing becomes fast and shallow. Panic attacks are often accompanied by a fear of being unable to breathe, and the fear of suffocation that follows exacerbates the attack. In short, both normal and pathological breathing are complex phenomena, with several levels of expression and a multifactorial physiological, psychological, and spiritual/cultural underpinning.

Phenomenologically speaking, breathlessness is remarkable in two intertwined ways: it is an overpowering sensation, to which we are deeply sensitive, but it is also behaviourally subtle, and so often invisible to others. Even we ourselves are not always aware of when we start to become breathless and why. Techniques such as meditation and mindfulness training help practitioners become more aware of their breathing, but this is hard to do when a person is immersed in an activity and her attention is turned away from herself. This phenomenon has led Gysels and Higginson (2008) to coin the term 'invisible disability'. The Janus-faced duality of breathlessness—the fact that it is so real and overwhelming to the person experiencing it and yet so invisible to those around her (and in particular to health professionals) merits close analysis. I will conduct the analysis using the phenomenological framework developed in earlier chapters. I suggest that breathlessness is phenomenologically salient for the reasons given above, and that a phenomenological analysis may reveal aspects of this symptom that have hitherto been unrecognized or have been obscured by a physiological approach to the symptom (compare Carel, Macnaughton, and Dodd 2015).

5.2 Phenomenology of Breathlessness

Trapped. That is what breathlessness feels like. Trapped in the web of uncertainty, bodily doubt, practical obstacles, and fear. The deepest fear you can think of. The fear of suffocation, of being unable to breathe, the fear of collapsing, desaturated to the point of respiratory failure. Even if illness descends upon you gradually, over many years, there comes a point—probably around the '30 per cent of predicted' mark—where high spirits and a positive attitude just can't cut it anymore.

You are faced with the shrinking of your world, choices, freedom, and eventually, your spirit. Many illness narratives take on a kind of dualist flavour—the body fails, but the spirit flourishes; the body is tethered to its failing organs, but the spirit rises free. Not true. Our embodiment determines our possibilities and delineates with extreme clarity what one is and is not permitted to do and be (as discussed in Chapter 3). The spirit is tethered to the body and its limitations cut deep into spiritual life. Transgressions are punished harshly. Push yourself too hard and you will pay: with nausea, with dizziness, by passing out, collapsing in front of strangers, or worse, in front of loved ones. Push yourself too hard and you will find yourself at the limit, or even beyond what you thought was possible, with saturation dipping below 70 per cent and your brain coming to a standstill. Your being imploding silently, fearfully, while your pants need changing.

Perhaps you have never been so breathless. If you are healthy, you probably haven't. So no, it is not like running for the bus; it is not like hiking in high altitude; it's more like what I imagine dying is like. One aspect of such breathlessness is aptly termed by medics 'air hunger'—the air is rushing in and out, but the reduced surface area of the lungs means that the oxygen isn't coming in fast enough and the CO_2 isn't removed fast enough. The result is the worst sensation I have ever felt and one that is very hard to describe, unless you've held your breath for far too long and feel your chest is about to explode. But you cannot raise your head above the water and take a deep and gratifying breath. All you can do is gasp and pant and hope that eventually the breathlessness will recede, even a little, and a glimmer of hope of recovery will emerge from the dense clouds of dyspnoea.

What is the phenomenology of this total sensation? Like pain, you cannot ignore it under any circumstances. You might try to pretend outwardly, but on the inside, the sense of loss of control—and the secondary fear of that sensation—will keep you tethered to your oxygen tank, and reluctant to go anywhere where your oxygen arrangements might be compromised. A vicious circle is created, where acute breathlessness must be avoided, thus leading to restricted activity, which in turns causes further deconditioning, which will cause the breathlessness to increase as fitness reduces.

The psychological impact is enormous. The sense of bodily doubt and insecurity gives rise to a host of psychological reactions: despair, fearfulness, anxiety, depression, loss of hope. Respiratory illness is a life sentence with no reprieve. No amount of good behaviour will matter;

you are chained for life to damaged, hyperinflated, lungs, to slashed functionality, and to the cascade of other losses that accompany this bondage.

The world shrinks and becomes hostile. The sense of possibility that accompanies objects disappears. A bicycle is not an invitation for an afternoon of fresh air and freedom. It is a relic of days bygone. Hiking boots now sit leaden in a cupboard. They are no longer 'something to be worn when going for a hike'; they have long been too heavy and hiking too hard. The inviting smell of mud and hills has faded from their soles, but you cannot find it in your heart to give them away. That would be admitting the finality and irreversibility of your condition (see discussion of Heidegger's tool analysis and of the transparency of the body in Chapter 2).

The physicality of every action needs to be calculated, considered, configured to suit your body's limitations. You now sit on the floor to receive the slobbery kisses of your toddler. Groceries have to be judged by their weight: a pint of milk—yes; four pints—no. Strawberries—yes; potatoes will have to wait for the online order. Strolling along on a beautiful summer day is censored by the gradient, amount of oxygen left in the tank, temperature, and fatigue. Everything becomes potentially debilitating, frustrating; a *problem*. A short walk to the library with a few books can turn into a forty-minute ordeal, accompanied by a panic attack.

Fun things take too much effort; your text 'sorry, I'm too tired to . . .' becomes a staple apology, cop-out, and truth. At some point, the lengthy deliberation about *whether* I am able to do something becomes so tiring so that falling into 'being unable' becomes second nature (see Chapter 3). This account is not intended as an exercise in self-pity. Rather, it is a realistic reassessment of what has been lost, what is no longer possible, what has turned from 'I can' to 'I no longer can'.

This phenomenology may read like a lament because illness is, in a way, an ongoing lamentation for things lost, gone, given up on. Life becomes a set of constraints, a levy charged at every twist and turn, making day trips expensive, outings exhausting, travelling not feasible, spontaneity overridden by physical constraints.

And as you give up, you take your loved ones down with you. We all fall prey to the shrinking, tight, ill-fitting world of respiratory illness. My heart goes out to my husband, as he sits poring over OS maps looking for flat ridges and riverside walks for a family outing. We reminisce about a game of tennis once played, long ago in Australia, and it seems more fiction than memory. I held a tennis racket and ran around a court—or did I? The impossibility of that gives the memory the quality of fantasy. I am now awed by people playing sports; even the everyday sight of people walking briskly up a hill impresses me now, in a convoluted, childish way.

But it is not a world without redemption. Illness can have an edifying effect on the ill person and those around her (Kidd 2012), as well as leaving room for well-being, enjoyment, and positive experiences (Carel 2013a). However, the redemption has to be underpinned by an existing resilience that may be absent, as well as fuelled by enormous amounts of optimism, creativity, and problem-solving—and the resources for these become depleted over time (compare Carel 2012). Occasionally I can raise my head above the breathlessness and try to enjoy the walk. But most of the time that enjoyment is buried under constraints and fears that fuel a sense of fearfulness and helplessness. Enjoyment is muffled by the constant breathlessness and limitation, and this muffled enjoyment becomes the new norm.

The constant interplay between desire and limitation destroys spontaneity in both individual actions, as well as in broader activities. This nested set of 'inabilities to be' mirrors Heidegger's analysis of purposeful action (see Chapter 3). Under normal circumstances, I type on the computer in order to write a report; I write the report in order to give feedback to a student; I give feedback to a student in order to fulfil my role as a philosophy teacher, and so on. Compare this nested set with the one of the breathless person: I walk down the road in order to get to my meeting with my student. But on the way I fall prey to a panic attack or coughing fit. I am then late for the meeting, so I cannot fulfil my role of being a philosophy teacher. I fail to be a good teacher, and so am unable to continue my purposeful activity.

The chain of nested purposeful activities is broken in illness. I am left alone with my inability to be. I am still embedded within the world of agency and action, but am unable to participate in it as I wish, as others might hope, and as a bodily prehistory, dating back to bodily certainty, sets me up to expect. The purpose of my actions is stripped away from them, as they become more idiosyncratic, out of step with others' actions, and more contrived. In this they also lose their shareability. I need to do things a certain way, at a certain speed, which is out of step with most others' ways. I am impotent and alone: frustration and alienation shape my experience.

5.3 The Geography of Breathlessness

Breathlessness creates a new world, a new terrain to be navigated. Where freedom and obliviousness once ruled, hesitation and limitation now dictate my movements. Holiday destinations have to be selected carefully; any trip beyond Europe demands months of stressful preparation. Distances have increased and I observe people walking several miles to work and back with fascination and envy. That I could once do the same is now a distant memory.

The gradient of any place becomes the most important factor. When invited to speak somewhere, my first question is always: is it flat? I approve of London, Oxford, and Helsinki; I am wary of Sheffield, Durham, and Bristol. I try to find ways of walking downhill without having to climb back up again. I sit in the car as my husband and son run up a hilly trail, and wave to me from the top. My perception of distances, level of difficulty, energy reserves, and level of effort required for an activity have changed. Everything seems further away and heavier than it was. Not only has my physical interaction with the terrain changed, but also my perception of it. A small incline now seems steep and daunting. What I once called a minor hill I now think of as a mountain. The feasibility of routine activities has changed. Walking to the shops and back is out of the question. The geography of my world has changed (see Chapter 3).

Travelling is almost beyond my reach. Getting on an airplane exacts a heavy toll in terms of financial, bureaucratic, and emotional cost. I prefer to spend Christmas at home than contemplate flying somewhere warm. Locating a flattish holiday destination, to which we can fly a relatively short flight, then getting quotes for specialist travel insurance, ordering the oxygen equipment in a foreign language (which is, without fail, a game of Chinese whispers), getting hold of the oxygen equipment I need for the flight, and then jumping through a string of bureaucratic hoops (pay £20 to see the doctor, get a doctor to sign the form, fax the form back to the airline, ring the airline to confirm, ring their assistance line at £1 per minute . . .). The tedium is insufferable but the anxiety around the oxygen delivery and the flight is downright disabling.

The question remains: what part of the experience of breathlessness could be mediated or alleviated by intervention, and which is intrinsic to the condition? Some of the difficulties described above are due to clunky bureaucratic arrangements, inflexible or lengthy processes, rather than the experience of breathlessness itself. That is why it is important to ensure that services such as the ambulatory oxygen service have plenty of ongoing input from patients and that such services are codesigned using knowledge and input from service users. The impact of illness upon one's life is determined by the convergence of a range of factors—bodily, physical, social, psychological, spiritual, and existential. Merleau-Ponty's remark that illness is a 'way of being' can be translated into the claim that the world as we inherit and organize it is less hospitable to certain 'ways of being'. Illness might change our perception of the world, but that world can, itself, do more or less to affect the sorts of perceptions that it gives rise to.

5.4 Respiratory Illness, Disruption, and Intelligibility

Meaning structures are destabilized or may even break down in illness, especially around the time of diagnosis or disease exacerbation. The time of diagnosis, especially in the case of a rare disease, can be toxic, tearing apart taken-for-granted expectations and goals. A mix of fear, confusion, anxiety, and a sense of unreality enfolded me when I was diagnosed. It was not just a nightmare coming true; it was also the most destabilizing event I have ever experienced (Carel 2015). I experienced my body as a fickle foe, a traitor, a disappointment; I was plagued by bodily doubt (see Chapter 4).

At the time, my physical condition was not too bad. But I still inhabited my body with fear and suspicion. Its fragility was revealed to me in all its forcefulness and the myriad of risks and dangers engulfed me: my lungs could collapse ('I'm surprised they haven't collapsed yet', remarked a thoughtful clinician). What does that mean? What do I do if they collapse? What does 'collapse' mean? How can lungs collapse? I didn't know the first thing about lung physiology. My lungs just did their thing. But now they were the object of fascination and learning. 'Come see this!'—'how unusual', I heard young physicians enthusiastically exclaim as they stood in front of my chest X-ray at the local hospital. Slowly, I learned about lung physiology and the disease process that was rendering mine unusable. Ah, I thought. So that's how they work. And oh, that's how they don't work. The image of LAM cells decimating my lung tissue at an accelerating rate of destruction horrified me. This was happening inside me and I had no way to stop it.

I had to rebuild. I had to ask what mattered, what still worked, how I could go on (but how could I not go on? That wasn't clear to me either). I regained a shaky foothold on the everyday. But then the everyday itself began shifting underneath me. Nothing worked anymore. Nothing was like before. The experience was of bodily change but also of the environment becoming less welcoming, less familiar (see Chapter 3). One day, I had to get off my bike and push it up a minor hill. Since then, whenever I drive up that road a small tremor still goes down my spine, like the spot of an accident that cannot be forgotten. This is where the early realization that something was deeply wrong met grim, clenched-jaw denial. I just need to cycle more, get fitter, I said to myself. I refused to acknowledge that cycling was no longer a possibility for me.

Later on I realized I couldn't chew gum anymore. It interfered with my breathing. Then, I tried to carry a bag and realized it was too heavy. A hike on the beach, which included scrambling on rocks, sent me to the ground, trembling with oxygen desaturation. Things I loved had to go. Cycling, gym classes, hiking, dancing, many forms of physical expression—all became impossible. Even minor things like walking briskly down the corridor, lecturing while standing up, or lifting my son out of the bath, became difficult. The habits I had—taking stairs two at a time, bending down to pick something up, running for the phone—had to be given up after being confronted with the level of breathlessness they caused. Within the space of a few months, during which my lung function rapidly declined, a life was destroyed and another was born. A sedate and sedentary life. A life of envy, of counting and recounting lost opportunities, given up activities, and of continuous reworking of the boundaries of the possible (see Toombs's account of illness as loss in Chapter 2). Those boundaries shrank by the week, eventually reaching an uneasy *status quo* of stability balanced awkwardly against disability.

The life that was created out of the illness was maimed, small, fearful. A life of rescinded possibilities, censorship of options, rejection of invitations and offers. 'I can't make it, it's too far'. 'I can't join you, I'll hold everyone up'. 'I can't camp'; 'I can't fly'; 'I can't swim'; 'I can't join the fun run'. I can't—or 'I am no longer able to'—became the starting point for nearly everything. It was hard to hold on to what was still possible (teaching, writing, strolling along a flat path), because it was swamped by what was no longer possible. I reside in the world of 'no longer able to', the world of inability to be, or do, so many things. Heidegger's conception of human being as 'ability to be' had to be expanded, I felt, to include this kind of being, imperfect though it may be (see Chapter 3).

Not only were my habits, expectations, and abilities disrupted and then resumed at a slower pace and smaller scale, but also—more dramatically—embodied normalcy was disrupted. Every movement would start out at its usual, brisk pace and then come up against my body's limitations and grind to a halt. Old habits were discarded and new ones created. But this was not a smooth process. Coming up against the boundaries of one's body is a painful and distressing experience (see Chapters 3 and 4).

Having witnesses made it worse: observing the surprise in people's face give way to pity only exacerbated the sense of abnormality. A few times I genuinely doubted that I'd be able to catch my breath because I've overdone it so badly. Hesitation now characterizes bodily movement, and the sense of bodily doubt hovers over the simplest of motions, ready to descend in the form of panic—what is happening to my body? Bodily certainty was replaced by bodily doubt. My body became a dreaded, cumbersome enemy, sabotaging my everyday, ruining my plans through a chest infection, flattening me into a bed for three weeks. I am in charge, it taught me, not you. The sense of control so pleasantly distinctive of healthy embodiment was exposed in all its falseness, as Toombs outlines (see Chapter 2). The body is a problem, an obstacle, a stranger. I never befriended that stranger.

I realized the extent to which meaning and intelligibility depend on patterns of embodiment. Once those were disrupted, the reconstruction work took years. I'm still reconstructing, still thinking how to reclaim the horizons of possibility and action that have been closed down. My habitual body was knocked back, surprised, and disappointed so many times it has learned, the hard way, to temper its desires. Habits crashed into, and then cracked against the wall of respiratory limits; most of these bodily habits are now defunct. They were replaced by new habits, more akin to those of the elderly. I now readily accept help, which I used to ardently refuse. Motor intentionality is now tentative. Movements are slow, hesitant. Many actions are contemplated but discarded as impossible or too difficult. As a result, my connection to the world—my intentional arc—has slackened, as Merleau-Ponty aptly puts it (2012). It is not the same intentional arc; it is not the same world (see Chapters 2 and 3).

Respiratory illness can be perceived as a series of losses, which the ill person, but also those close to her, incur. These include the lossess described by Toombs (1987). The loss of wholeness arises from bodily impairment leading to a profound sense of loss of bodily integrity. Bodily doubt, which was described in detail in Chapter 4, articulates this loss. The body's pre-reflective sense of wholeness is naïve in two senses. First, it is whole, unbroken, undisrupted. The unity of motion and action, perception and execution, is largely unspoilt prior to illness. Second, we have the innocent sense that nothing could go wrong, expecting tomorrow to be like today, today like yesterday. This inductive sense of continuity is irrevocably lost in illness. The awareness and anticipation of the loss of this sense is itself a source of suffering and distress. One only becomes really aware of a tacit bodily or worldly confidence once it has been challenged, broken, or lost (see Chapter 4). Interestingly, this is a case where knowledge limits the ill person, robs her of her innocence, and constrains, rather than liberates and enriches.

The loss of certainty could not be more explicit. The expectations and past experiences on which present certainty is scaffolded become irrelevant, and are suddenly perceived to be illusory, false, or even cruel. Bodily change brings about a change to other foundational aspects of selfhood: beliefs, judgements, trust, hope, and desire are all torn up and then either replaced or given up on. Once again the intimate link between body and mind becomes central: one cannot lose bodily certainty without that doubt bleeding into all other life domains.

Loss of control causes immense anxiety. Knowing that your body can—and will—betray you, but not exactly when and how, constitutes a sense of loss of control. Breathlessness might be more or less predictable, but being able to plan a year ahead for a conference or holiday becomes impossible: what if I have a chest infection? What if my lungs collapse? I have introduced a winter embargo on travel—the risk of a cold turning into a chest infection and keeping me in bed for weeks is too high. Travel exacerbates the sense of loss of control. What if my train is cancelled? What if there's a storm and I'm stranded somewhere with no oxygen? The sense of loss of control can only just be kept at bay when everything is going fine. But a weather event or my husband falling ill is enough for loss of control to take over. The loss of control is so pervasive that only the most sedentary and pre-planned lifestyle can accommodate it, barely covering its ugly nakedness.

The loss of freedom to act is experienced as the terrifying shrinking of one's world. It intertwines creepily with the loss of the familiar world, as activities and actions become barred, impossible, and the familiar world must be reconstructed from fake, cheap replacement materials. Like a temporary false tooth, closing a gaping hole in one's mouth, this replacement world is anchored in a painful sense of wanting to inhabit another world, wanting another life, another body. The impossibility of that is cemented by the loss of freedom to act. Suddenly there are more things I am unable to do than previously. As film director David Cronenberg says: 'first we grow up, then we grow down. There's only a moment where there are a few years of the illusion of stability. It doesn't last long' (quoted in Rodley 1992, p. 124). We do our best to cultivate the illusion of stability, but the 'growing down' of illness and ageing is humiliating. It hurts.

Illness shapes all you do, and possibly who you are. It shapes how you are perceived by others and your own image of yourself as a social being (Sartre 2003). I have long given up trying to address, or even process, the looks I get, as well as their hidden, and not so hidden, meanings. I am much less social. I walk in my own groove, oblivious to others' stares, whispers, awkwardness. I have to soldier on, because we see strangers on the street every day. And the aftershock of internalizing the stares and whispers has to be contained, triaged, suppressed. I feel more alone now because others cannot understand the contours of my world. I no longer walk in their tracks, share their freedom, take for granted the commonality of health, its hegemony, its normalizing force.

5.5 Objective vs Lived Body

A well-known enigma in respiratory medicine is the discrepancy between objective lung function, as measured by lung function tests, and subjective feeling and function. Patients often over- or under-perform, when correlating their everyday function with their objective lung function. This is the mismatch between objective and subjective measures in lung disease (Jones 2001). No one can explain or predict why one person is crushed by their respiratory limitations and soon becomes housebound, while another patient, with equally diminished lung function, continues to live fully and actively. Is it merely psychological resilience that enables this? Is the second patient more resourceful, finding ways to perform daily tasks more efficiently? Do they come up with shortcuts and adaptations that allow them this well-being? Or is the large variability in lung function (up to 20 per cent variation in normal daily fluctuation) simply preventing us from measuring the true capacity of a particular patient?

These questions take us back to Merleau-Ponty's distinction between the objective and lived body, discussed in Chapter 2. The oscillation between the lived, and more importantly the habitual, body characterizes the onset of illness. I set out to lift something or gesticulate generously. Each of these demands more oxygen than my lungs can deliver and within a few seconds, I am awash with the flush of desaturation, imminent dizziness, and the creeping sensation of loss of control. I now adjust, slow down. But initially, when I was just diagnosed, I overdid it nearly all the time and was punished harshly by my body for my transgressions. The threat of losing control and the odd sensation of bodily betrayal, accompanied by a sense of bewilderment—*why* is my body doing that?—characterized the early period of my illness.

Later on my bodily habits changed. Brisk movements were censored and the interplay between the habitual and objective body minimized as my bodily style was entirely changed by bodily limitations. Slow, regal movements; not even thinking about joining in an autumn leaf-throwing chase in the park; trudging slowly uphill in the pouring rain, stopping every few steps to catch my breath while getting soaked; never trying to walk and talk at the same time—these became my new bodily habits. It was as if life was now lived in slow motion. I envied people to whom briskness was never a habit—they probably didn't have such a wide gap to close. I had to cross this gap and reposition myself firmly on the other side. Slow as a sloth, minimal in my talking, tiring easily, counting pavement stones to pass the time while catching my breath—I live in slow motion. The loss of bodily ability and energy soon translated into a more social fatigue. I was unable to socialize as much, often too tired to go out, put off by the practical hurdles involved.

The loss of transparency of the objective body and the tense relation between it and the lived habitual body came to the fore (see Chapter 2). The act of breathing and the modulation of breath became explicit, primary tasks. I have to continually adjust the flow rate on my oxygen tank, depending on what I'm doing. I'm continuously tethered to this flask. One of my hands is often not free, as it's either holding or adjusting the oxygen. All this artificial and conscious engagement with my breathing apparatus, both biological and mechanical, stands in sharp contrast to the natural tacit way in which we normally modulate breathing, increasing the rate of gas exchange when going up stairs, for example and then slowing it down, without noticing it. That's the beauty of it: it happens by itself. You don't need to worry about it. You don't need to do anything. It just happens by itself. However, for the respiratory patient, the explicitness of the process, the constant monitoring of the oxygen flow rate and level, becomes second nature and limits movements and capabilities in a way that itself become part of the new habitual body—the breathless one.

This loss of spontaneity ties itself to a spatiality of illness that is much different to the healthy one. Initially, I experienced new challenges as novelty—I would stop short, thinking 'huh, I can't lift this anymore'. Later on the novelty was replaced by censorship and that, in turn, by a new set of bodily habits, which invariably impact on other habits as well. Old habits were quickly forgotten. A new way of being—a breathless way—has become the norm, further distancing me from healthy people whose norms radically differ from mine. When I walk downhill I wonder gravely—how will I ever get back up? An inability to be has become a way of being.

5.6   Invisibly Breathless

I recently helped a friend who has the same respiratory disease as me (lymphangioleiomyomatosis, or LAM). She is very ill and relies on ambulatory oxygen, a walker/seat, and an oxymeter to check her oxygen saturation levels. For her, every step requires enormous respiratory effort. She is unable to walk more than a few metres and her recovery is very slow. I walked with her, stopping and waiting for her once we arrived at the doorstep of the house we were going to, about twenty metres from the taxi. Once we reached the doorway, there were two steps to be negotiated. And once past the threshold, several minutes of recovery were needed by my friend in order to be able to enter the house.

Watching her and walking alongside her, I was trying to think of ways to be helpful by drawing on my own experiences of breathlessness. I remembered that the last thing you want to do is talk (requires lots of air!); so having to answer questions or explain how you feel are serious hindrances when you are trying to stay upright, regain a decent level of oxygen saturation, and calm your panting down. First rule: don't make conversation. I tried to keep quiet and just be present and patient. However, that can be awkward—I was standing there, observing what seemed like acute distress, while not knowing what I can do to help. Second rule: don't stare at the breathless person; better yet—leave the room and give them privacy during recovery.

Throughout the evening I did small jobs for my friend: refilled her glass, cleared her plate to the kitchen, fetched her bag, and rummaged in it for the various items she needed. I checked the oxygen cylinders and reassured her that there was plenty left. I realized that I don't know what tasks she would like help with and which she would like to do on her own. I learned that even the act of looking for her phone in her bag was strenuous for her, and something I could do. Third rule: don't assume the breathless person can or cannot do anything; ask them if they'd like help with even the most trivial task. An outsider cannot know what is difficult for her.

Throughout the time I was watching my friend, I was thinking how incommunicable breathlessness is. The distress and sense of impending suffocation, the panic bubbling up in severe breathlessness, the sense of loss of control, are entirely internal, impenetrable, as invisible as the oxygen the respiratory patient's body so desperately craves. An observer can see the person standing still and panting, the laboured breathing— but these do not really convey the subjective sensation of severe distress.

The experience of breathlessness looks so innocuous from the outside that conveying the nature of that experience is extremely difficult and has not yet been philosophically analysed.[3] The experience of illness remains largely invisible and difficult to communicate, and this for several reasons. First, in illness the norms of feeling and action shift considerably to create a different spectrum. For example, the sensation of pleasant breathlessness or exertion, that can be thoroughly enjoyable when, say, dancing or running, entirely disappears from the spectrum of breathing sensations in respiratory illness. There are only two (somewhat graded)

---

[3] This task is undertaken in *The Life of Breath* project, mentioned above. The project aims to conceptualize the experience of breathlessness in ways that will be useful in the clinic.

sensations: one of unstrained breathing (e.g. in rest) and the other of unpleasant unrewarded breathing (e.g. when walking, lifting, or bending). The shift from one to the other is sudden and the recovery back to unstrained breathing is in itself uncomfortable.

The broad spectrum and gradual transition from one state to the other that is experienced in healthy breathlessness is compressed into a much narrower experiential range. There is either no movement and no breathlessness, or some movement and with it the inevitable, unpleasant breathlessness. The spectrum becomes a binary: one is either breathless or not. There is no graded transition from breathing at rest, followed by somewhat deeper breathing when lifting something, moving onto more effortful breathing when cycling or dancing, and reaching panting when exerting oneself at 100 per cent effort. Within that range, it is normal to experience some exertion and some breathlessness as pleasurable. Think of the warm sensation of dancing, cycling on the flat, or walking briskly in the countryside. These are all experiences of exertion, but of a pleasant nature. One way to differentiate normal from pathological breathlessness is that pathological breathlessness does not have this pleasant quality. It is negative and threatening.

When the norms shift and normal and pathological experiences diverge, communication suffers. The ability of a healthy person to understand the constricted experiential space of the breathless person is limited because of the lack of a shared experiential background. The shared norms of enjoyable exertion, normal functionality, and their accompanying world, disappears. Walking to the pub becomes a lonely affair, reinforced by the constant inner discourse about breathing ('shall I cross diagonally, to make the gradient less severe?', 'do I need to put up the oxygen flow?'). The company of others only serves to reinforce the loneliness of walking for the respiratory patient. The social glue of 'being-with' is removed from the human activity of walking together, usually so solicitous. Instead, it becomes an embarrassing affair, the respiratory patient's deficiency accentuated by the artificial slowing down and awkwardness of the healthy walker (see Chapter 3).

Second, the discrepancy between the perceived ease of a task (say, picking up a pen from the floor) and the great challenge it poses for the ill person creates a doubt in the mind of the observer. Not a doubt about the disingenuousness of the patient, but a more philosophical doubt about the shared understanding of norms. What is 'far' or 'just round the corner'? What is 'reached easily' and 'not heavy'?[4] Once confronted with the widely differing perspective of the respiratory patient, the nature of the task itself has to be reconfigured. The complete certainty of breathlessness as experienced subjectively, and the complete doubt it generates when viewed from the outside make it akin to pain. As Elaine Scarry writes:

Thus when one speaks about 'one's own physical pain' and about 'another person's physical pain,' one might almost appear to be speaking about two distinct orders of events. For the person whose pain it is, it is 'effortlessly' grasped (that is, even with the most heroic effort it cannot *not* be grasped); while for the person outside the sufferer's body, what is 'effortless' is *not* grasping it (it is easy to remain wholly unaware of its existence . . .). So, for the person in pain, so incontestably and unnegotiably present is it that 'having pain' may come to be thought of as the most vibrant example of what it is to 'have certainty,' while for the other person it is so elusive that 'hearing about pain' may exist as the primary model of what it is 'to have doubt.' Thus pain comes unsharably into our midst as at once that which cannot be denied and that which cannot be confirmed (Scarry 1985, p. 4).

Pain is famously conceptually and medically complex and a standard philosophical example of scepticism about other people's internal states. The sceptic's challenge is an epistemic one: when I see someone writhing in pain, I can never know, in principle, if they are really feeling pain or merely behaving as if they do. Of course normally when one sees someone in pain, the immediate judgement is that they *are* in pain; one only adjusts this belief if it becomes clear that they are feigning the behaviour. Moral 'sentimentalists' make this point, and claim that our natural moral sentiment is to accept and sympathize with the pain of others, not to automatically admit the possibility of deception. However, the sceptical challenge remains intact, and perhaps the epistemic gap Scarry points to is what allows the unshareability of pain and the sceptical challenge to persist in the face of ordinary experience.

I suggest that we could say something similar about the unshareability of breathlessness. There is a body of evidence that points to this unshareability as undermining the quality of clinical care and knowledge

---

[4] There is also a sense that it is unfair that our inherited norms and values are structured around health, when so many people are, or will inevitably become, ill. Since it is the ill or disabled who generally require more consideration and accommodation, that is doubly unfair. I am grateful to Ian James Kidd for raising this point.

about breathlessness (Johnson et al. 2014). Despite being so common, breathlessness remains invisible, opaque, or refractory. This opacity poses a particular challenge to health professionals, as the subjective experience of symptoms like pain and breathlessness is largely invisible.

5.7 The Social Architecture of Breathlessness: Sartre's Three Orders of the Body

So far this chapter has discussed one's subjective experience, and how difficult it is for that experience to be understood by another. But being viewed by another, as Sartre shows, allows for a third perception: a self-perception of myself *as perceived by others*. Sartre refers to emotions such as embarrassment and shame as ones that express to us most vividly our existence as social beings. When breathlessness is experienced in front of others the sharp awareness of how I appear before another comes to the fore. This is so in many cases of stigma, as noted by Goffman (1963).

Those suffering from pulmonary dysfunction resort to many tricks to mask their breathlessness. When they stop to catch their breath they might pretend to be looking at the view or checking their phone. They feign disinterest in social activities they feel unable to participate in. They need to plan with great care any outing; what may seem like an easy and spontaneous decision to go for a stroll or meet in the pub, may be the result of hidden planning. And then there's the oxygen tank and tubes. Children stop and stare, adults ask intrusive questions, teenagers smirk.

The social architecture of breathlessness is not without its fault lines and cracks. It eats into chance encounters and exposes the breathless person: you are not only on show but also trailed by tacit assumptions and guesses—is it because you smoked? Is it contagious? How long will you live? The questions linger, unvoiced, in the silent curiosity impregnating the air wherever the oxygen-user walks. Mixed with pity, terror, and denial, the curiosity can become lethal if you let it. It is a hard lesson to learn.

If we return to Sartre's three orders of the body, discussed in Chapter 2, we can see how these play out in the reality of life with a respiratory illness. The respiratory patient perceives her own difficulties in the first person. She is also perceived by others as a physical entity with particular features: in this case, disabled. She is then perceived by herself as seen through the eyes of another. It is this third order of the body that is so poignantly described in illness narratives. Feeling self-conscious when undergoing a physical examination by a health professional is an example of this third order (compare Carel and Macnaughton 2012). In this order I experience my body as it is seen by another, no longer with the pre-reflective ease with which I previously experienced it.

The coexistence of these two perspectives is a prominent feature of the social experience of illness. In that context, the body is experienced as a burden, a problem, but these feelings are often exacerbated by the social experience of my body as an aberration, an affront to others. My body is no longer 'my body as I exist it', but is now doubly problematized: once in my lived experience of my body and again in my experience of that body as perceived by others. Breathlessness is therefore often compounded by embarrassment when experienced in public. The subjective sense of discomfort is replaced by a sense of shame of being exposed in my breathlessness and inability to another.

In this interaction I bring other perspectives to bear on my own subjective stance. The realization that these other points of view perceive me in my inability to be becomes part of that experience, or of breathlessness, as in the case discussed here. Perhaps the third order of the body points to a dynamic in which not only are we objectified, but also it may be that others desire to objectify us in a way reflective of certain vices— e.g. unkindness, inconsiderateness, or inability to empathize. It is not arbitrary that Sartre's examples of the third order of the body are all negative: shame, shyness, and self-alienation are paradigm examples of this intersubjective dynamic. Reflecting on this dynamic morally, we may say that there is something vicious in those thoughtless stares and objectifying gazes that aim to protect the one who gazes from the empathic understanding that she could have been (and is in principle) the one being gazed at and objectified.

5.8   Ill, but Well

As a preamble for the next chapter, I would like to raise three interrelated points on the possibility of well-being within illness, and within respiratory illness in particular. Throughout the chapter I have articulated an element of loss and lamentation in illness. However, illness also has unintended and unexpected positive aspects to it that are poorly understood and rarely documented within medicine while being of the utmost importance to patients (see Chapter 6). The role of a phenomenology of illness is to account for the richness and diversity of the illness experience, and thus also to articulate the positive, often unexpected, consequences of illness.

These positive consequences include improving personal relationships and increased intimacy with family members and friends; a sense of purpose and focus; rediscovering the self; resilience; pride arising from overcoming difficulties; and more generally what Jonathan Haidt calls post-traumatic growth (2006). The effects of illness are thus much more diverse and modulated than might seem on first blush. As we will see in the next chapter, the positive effects of illness and the overall stability in reported levels of happiness in patients are both striking and worthy of articulation. Thus, another role of a phenomenology of illness is to articulate the possibility of well-being within illness.

However, traumatic events do not always lead to post-traumatic growth. People who have suffered trauma often describe a world that has been permanently shattered. Some cases of illness will cause irreparable damage, pain, and suffering that will not be a source of growth, but of permanent disablement and distrust. Some cases of illness will be of the kind described by Arthur Frank (2010) as 'restitution narratives', in which the endpoint of the narrative is one in which the ill person is somehow better off than she was prior to the illness, even if her health has not been restored. Other cases will be too chaotic, painful, or traumatic to form a narrative at all. The violence of illness, its eruption and destructive force cannot be understated; the ideas set out below about the positive consequences of illness are in no way meant to belittle this. With this caveat in mind, let us now turn to the notion of well-being within illness.

First, the possibility of well-being within illness is notable and although it is hard to perceive from the outside, it can be both achievable and significant to the ill person herself. As will be discussed in the next chapter, viewing illness from the outside precludes seeing the complexity and variation in the experience, and outsiders often fall prey to the 'focusing illusion' (compare Schkade and Kahneman 1998). When viewing it from the outside, we tend to reduce the experience of illness to a monolithic and clichéd view, due to lack of information and failure of imagination and memory (Gilbert 2006). In fact, the experience of illness is rich, changeable, and has a broad bivalent spectrum of both good and bad.

It is important to remember, when confronting illness or treating patients, that it is possible to experience periods of stability and well-being within the context of chronic illness, and these periods are important to articulate, cultivate, and theorize. For example, the shifting perspectives model (Paterson 2001) provides a framework within which to understand how well-being is possible even within the context of severe illness and disability. This model proposes a view of illness as a continually shifting process in which an illness-in-the-foreground or wellness-in-the-foreground perspective have specific functions in the person's world (2001). Illness comes to the foreground during times of diagnosis, symptom exacerbation, or disease progression, but may remain in the background during other times. It is during those times that well-being, hope, enjoyment, and positive experiences can be sustained.

Second, a phenomenology of illness itself can contribute positively to one's experience of illness. Experiences that are initially chaotic and confusing can, once phenomenologically articulated and ordered, lend new structure and order to the ill person's experience. This articulation is one example of the positive force of a phenomenological process of ordering and reflecting on illness experiences. The articulation itself can be a positive process (Carel 2012). The ways in which we conceive of and describe illness in turn shape how we experience and respond to it. Both the conception and narration of illness is interpersonally negotiated, rather than done by the ill person in abstraction from the interpersonal, the social, and the cultural. We thus have in phenomenology a robust tool for a joint venture of shared description, reflection, and articulation of the experience of illness. I suggest that the process is in itself helpful.

This use of phenomenology, for example in the form of a phenomenological patient toolkit (Carel 2012) provides much-needed structure and meaning to the flux of experiences, as well as providing a first-person confirmation of the value of phenomenological analysis. A philosophical analysis of changes in the structure of experience during illness can provide a theoretical explanation of certain natural responses—such as fear, anger, frustration—in a way that restores a sense of intellectual control of those responses, as well as providing coherence and a vantage point from which to view and process such responses.

Finally, patients, health professionals, carers, and relatives of those who are ill have responded to phenomenological work by saying how positive and useful they found this framework, because it gave clarity and legitimization to their own experiences and has thus lessened their sense of isolation and of being unheard. Work on epistemic injustice in healthcare (Carel and Kidd 2014; Carel and Györffy 2014) has met with similar responses and is yet to be harnessed more fully to improve patients' experience and to increase understanding of sites of difficulty within the healthcare system (for a full account of epistemic injustice in illness see Chapter 8). Articulation is a powerful tool in a number of contexts and theoretical frameworks, such as narrative medicine, phenomenology, and epistemic justice. The articulation in itself promotes well-being by being an enriching and reflective process in its own right, but also by being a springboard for increased self-understanding and better communication about illness.

5.9 Breath: Beginning and End

Breathlessness is not only a symptom; it is a constant horizon that frames the experience of a respiratory patient. It is a limiting factor, and a condition of possibility for any action or experience the ill person may have. It also has a rich symbolic and psychological meaning. Our first and last breath mark the beginning and end of life. A baby's first breath, noted by her cry, is a symbolic moment of joining humanity, with a voice propelled outwards by her tiny lungs. It is the breath of life. Here I am, she says. Hear me. Witness my efforts to exist and communicate. The last breath is rasping, irregular, forced. Or it can be barely perceptible. It says: I am running out of breath which is a running out of life. And in between the first inhalation and last exhalation lies all of life, continuously rising and falling, inhaling the exterior and exhaling the unneeded, taking in and letting out. This life can be lived within the horizon of normal breathing, but may be lived in pathological breathlessness.

The richness of the experience, and the way in which it differs from normal breathing have only been sketched in this chapter and are yet to be studied in detail. But the intimate entwining of life and breath makes breathlessness a juncture of the physiological, psychological, existential, spiritual, and cultural. It cannot be studied solely as a symptom.

It must be understood and framed in a multidisciplinary context. This chapter heralds the start of an attempt to fill a lacuna in the philosophical literature, which lacks a philosophical—and more specifically a phenomenological—analysis of breathing and breathlessness. I hope it demonstrates the richness of this experience as well as its potential contribution to a richer understanding of what are taken to be medical symptoms.

6

Is Well-being Possible in Illness?

Sweet are the uses of adversity,
Which like the toad, ugly and venomous,
Wears yet a precious jewel in his head

William Shakespeare, *As You Like It*,
Act II, Scene i, ll.12–14

Up to now, the discussion has focused on negative aspects of illness experience; for example, the shrinking of the ill person's world, bodily doubt, and loss. However, it is also important to acknowledge the positive aspects of this experience. This is not an attempt to deny the negative, but a phenomenological effort to illuminate the hidden and often surprising positive effects illness may have on the ill person and those around her. In this chapter I suggest that illness is, of course, a distressing experience, but that this is compatible with the claim that it often does not, and need not, have a detrimental effect upon well-being.

Not all types of illness experience and not all symptoms permit well-being. Chronic pain is one such example, as is incontinence. Both have been noted to have a dramatic effect on patients' levels of well-being (Angner et al. 2009). In this chapter I explore the significance of the existing evidence that many medical conditions cause little diminishment in well-being. These are medical conditions that can be well controlled and cause relatively manageable problems, such as high blood pressure, asthma, or diabetes. Since patients may live with these conditions for many years, a process of adaption has time to unfold. Another set of cases in which well-being is not much diminished include some kinds of bodily differences and disabilities, which can also be adapted to (for example, paraplegia or vision loss).

The question raised here is whether well-being can be maintained in cases where illness causes subjectively unpleasant experiences, such as pain, nausea, fatigue, dizziness, or confusion. Can well-being persist even in cases where one feels ill or poorly?[1] The surprising answer emerging from empirical work is that it can. This subset of cases will be the focus of this chapter.

The idea that happiness is compatible with illness is particularly implausible in the case of many, if not most, mental disorders, notably disorders that include a depressive element. Thus I will not be discussing mental disorder in this chapter and will focus my argument here on somatic illness. Mental disorder in its very essence contains an element of suffering, described as 'problems in living', 'distress experienced by patient', 'harmful dysfunction', or of suffering itself being listed as a symptom, e.g. 'labile affect' or 'low mood' (Fulford et al. 2005; Wakefield 1992). If disturbed or low mood is a constituent symptom of a disorder, it makes little sense to ask whether well-being can be experienced in tandem with it.

Another change of tenor comes in this chapter. So far we have proceeded descriptively, using phenomenology to capture the general characteristics of illness. In this chapter, I move to an evaluative perspective, asking what is the impact of illness on well-being? Does illness necessarily herald a reduction in well-being, or might the relationship be more complex?

Serious illness is usually perceived as an insurmountable obstacle for a good life. Surely, one may think, one of the conditions for leading a good life (roughly conceived of as fulfilling and happy, although not exclusively so) is good health. How can one have the things that make up a good life—freedom, new experiences, joy—when one's life is encumbered by illness, limitation, and incapacitation? Illness is usually seen as bad luck or as a disaster that strikes an individual. This pathophobic conception of illness has a strong hold on our culture and imagination, but empirical qualitative evidence shows that illness can also be an opportunity and a challenge that bring about edification and personal growth. In this chapter I examine these questions and suggest that the reason we normally observe only negative aspects of illness and hence consider it a serious misfortune is that our ability to conceive of an ill life that is also happy is limited by our 'outsider's perspective' on illness.

---

[1] I am grateful to an anonymous reviewer for helping sharpen this question.

Long-term illness is often episodic in nature. It is also global in reach. The two characteristics are not mutually exclusive. Chronic illness is, as Merleau-Ponty described it, a 'complete form of existence, something that cannot be merely overcome or corrected for, but requires a holistic evaluation that incorporates adjustments made for lost abilities (2012, p. 110). As a form of life it contains within it all the familiar tensions and extremes we know from other such forms. Illness can be an overarching theme that can structure one's entire life. But it can also—and often does—recede into the background in a way unimaginable to the healthy outsider.

This chapter explores these two dimensions of chronic illness—its *global* and *fluctuating* nature—which capture this 'form of life'. I raise two questions: first, does illness affect one's well-being? I explore empirical evidence on the relationship between health and happiness and suggest that, surprisingly, the answer is: 'not as much as you'd expect'. I then offer several explanations for this finding. I examine first-person reports of ill people, to see what they say about their own happiness and lives. Such reports, inspired by a phenomenological approach, help understand how it is possible for ill health not to undermine well-being.

A second, related question is this: given that illness does not greatly reduce well-being, why do we conceive of it as one of the most terrifying events that can befall a person? I turn to recent work in empirical psychology to suggest some answers to this question. In the final section I suggest, following remarks by Julia Annas, that happiness is an achievement that requires thought, planning, and work, and that this view of happiness offers further support to the view that illness need not significantly affect long-term well-being. I conclude that illness provides us with a context and opportunity for the kinds of reflection that are the condition for happiness on Annas's account; in this respect her account of happiness provides philosophical grounding to the view that illness does not significantly affect well-being and also opens the door to viewing illness as philosophically productive, an issue explored in Chapter 9.

As we have seen in previous chapters, the experience of illness varies from one person to another, as well as depending on cultural context, social situation, historical framing, and other factors. Even for the same person, an experience of a symptom or aspect of illness may vary across time. Experiencing a symptom for the very first time is entirely different to experiencing it once diagnosis is made, or after many years of coexisting with a chronic condition. So how can we talk about 'the' (or even 'an') experience of illness or of disability? Surely that would be like talking about 'the' experience of love or of parenthood—how is it possible to generalize such an experience?

Because the experience of illness is so diverse and multidimensional, we need a descriptive method that does not try to subsume the richness and diversity of experience under predetermined conceptual categories. We also need a method that recognizes and values first-person reports and appreciates how crucial they are to understanding illness and disability. As we have seen, phenomenology offers an apt philosophical method for achieving just this.

There is another way in which phenomenology can illuminate the experience of illness; it can show us the limitations of the 'outsider's perspective' on illness. In other words, when we make societal and policy decisions, as well as personal decisions, we need to be acutely aware of the differences in judgements about whether a life with illness or disability can be happy or worthwhile, when made by those living with the conditions and those who do not. The difference between first- and third-person perspectives, discussed in the previous chapters, will now be used to answer the question why we often think that a life with illness or impairment is less happy than do those living with the condition. I am not claiming that a life with illness or impairment is necessarily happy, only that it need not be substantially less happy than a life of a healthy person.[2] In other words, health is neither necessary nor sufficient for happiness.

6.1 'Insider' and 'Outsider' Perspectives

I get into a taxi. It is late at night and I am tired; I have just come back from a conference. I carry with me an oxygen cylinder. The sound of the oxygen streaming through the tubes is clearly audible in the quiet night. The taxi driver is silent for a few moments, but I feel the inevitable question brewing. By the time we reach the second roundabout, he asks: 'what is that?' 'Oxygen', I reply and explain I have a chronic lung condition. He pauses for a moment and then says: 'I pity you'. It turns

---

[2] Indeed, no life is necessarily happy, since a person can enjoy all the goods of life (status, wealth, health, etc.) and still be miserable.

out that he is a devout Muslim and is well-intentioned; when I leave the taxi, he promises to pray for me. I am not offended by his offer of pity and prayer; I have a black belt in chronic illness, and have had many such (and much worse) exchanges before. But I think about this phrase, this sentiment: 'I pity you'. Pity is a common response to illness or disability: they appear to the outsider as terrible and insurmountable predicaments. Indeed, disability activist Harriet Johnson (2003), who suffered from a degenerative neuromuscular disease, wrote poignantly about how people approached her on the street to say 'if I had to live like you I think I would kill myself'.

However, as it turns out, disabled and chronically ill people do not have markedly elevated suicide levels, and seem to have only somewhat lower levels of well-being compared to healthy people (Gilbert 2006, p. 153).[3] Many studies report no difference in levels of reported well-being between groups of people with a variety of medical conditions and healthy controls. For example, Angner et al. (2009) studied 383 adults in the community, examining the relationship between subjective health (as assessed by the individual), objective health (as assessed by an objective measurement such as comorbidity count), and happiness (subjective well-being). They conclude that 'medical conditions are associated with lower happiness scores only if they disrupt daily functioning or are associated with social stigma', and give two such examples: pain and urinary incontinence (2009, p. 510).

Another study compared a group of haemodialysis patients with healthy controls and found that both groups overestimated the impact of haemodialysis on well-being. In fact, both groups reported a similar level of well-being (Riis et al. 2005, p. 6).[4] De Haes et al. find that 'Psychological symptoms do not automatically accompany physical

---

[3] It is important to note that well-being in this context denotes subjectively measured well-being, or level of happiness. For a period, discussions focused on whether well-being is subjective or objective, and how we should measure it. Should we measure some objectively observable feature, such as the amount of smiling or brain activity, or should we simply ask people how they feel? More recently many researchers in psychology and happiness studies agree that subjective measurement corresponds, roughly, to what we mean in everyday talk when we refer to happiness or well-being (Lyubomirsky 2007, p. 34). As Gilbert writes: 'the attentive person's honest, real-time report is an imperfect approximation of her subjective experience, but it is the only game in town' (2006, p. 70).

[4] Similar findings have been reported by Chwalisz et al. (1988), Chaung et al. (1989), and de Haes and van Knippenberg (1985).

distress' (1990, p. 1036). In a literature review of quality of life in cancer patients de Haes and van Knippenberg conclude that 'in general the results from comparative studies are meagre and do not support the assumption that cancer or cancer treatment lead to a significantly lower quality of life' (1985, p. 815). And finally, Brickman et al. report that paralyzed accident victims are only somewhat less happy than lottery winners (1978). Ubel et al. (2005) addresses this body of evidence by saying: 'People experiencing a wide range of illnesses and disabilities often report paradoxically high levels of [quality of life] and mood.'

These studies show that there is little correlation between objective health and well-being, although the UK's National Institute for Health and Care Excellence (NICE) clinical guideline 91 summarizes evidence for a moderate increase in depression levels, in particular in the first year following diagnosis (2009, p. 27). Some happiness studies researchers and health economists have found that happiness levels dip on diagnosis and symptom appearance, but that this effect disappears within about a year, during which happiness levels return to baseline (Angner et al. 2009; Lyubomirsky 2007, p. 50).

This can be explained by appealing to 'hedonic adaptation': we adapt to—and therefore cease to feel the impact of—changes to things that affect our hedonic state (e.g. the car we drive, size of our house, and even marital status) (Lyubomirsky 2007, p. 47ff.). Hedonic adaptation also works when the changes are negative, for example, falling ill, having to adjust to continuous medical treatment, and loss of mobility (Riis et al. 2005). As Angner et al. note, 'insofar as a medical condition affects happiness at all, it will only do so for a relatively short period of time after the diagnosis or the appearance of the symptoms' (2009, p. 509).[5]

Importantly, this holds for chronic medical conditions and disabilities, where long-term adaptation is possible. Of course in the case of acute illness we will not see such adaptation. A person who suffers violently

[5] More generally some findings show that overall life circumstances (including upbringing, marital status, income, health, etc.) account for only 10 per cent of our happiness (Lyubomirsky 2007, p. 20). This view, known as 'set-point theory', holds that the stable component of well-being is largely determined genetically (Lykken and Tellegen 1996). Set-point theory views the effects of life events as 'transitory fluctuations about a stable temperamental set point or trait that is characteristic of the individual' (1996, p. 189). Lykken and Tellegen go so far as to claim that the well-being of one's identical twin is a better predictor of one's self-rated happiness than one's own education, income, or status (p. 189).

for two weeks from the Ebola virus and then dies will not experience a return to previous levels of well-being. Nor can we imagine that undergoing this kind of violent illness will leave the Ebola sufferer's level of happiness intact.[6]

How can we explain this discrepancy between our powerful intuition that illness is a terrible misfortune and data showing that when we fall ill we adjust and regain our previous level of well-being relatively quickly? Two issues require explanation. First, why does illness impact on long-term well-being to a lesser degree than anticipated? And second, if illness does not make people unhappy, why do healthy people view illness as a purely negative and much-feared event?

One way of accounting for this tension is by attending to the difference between 'outsider' (third-person) assessments of ill health made by healthy people and 'insider' (first-person) assessments made by ill people. The difference is considerable and can be presented using the following example. In a study of European quality of life assessments, Dolan (1997) identified eighty-three states of illness that healthy interviewees rated as being 'worse than death', i.e. interviewees ranked these states lower than not being alive at all on a numerical scale. However, people who live with these conditions report only a slightly diminished level of well-being as their healthy counterparts and most prefer to go on living with the condition judged by outsiders as 'worse than death'. There is a gulf between how such states of illness are experienced first-hand and how they are perceived by observers.

Here are a few reasons why such a gulf exists. First, most healthy people have only sketchy ideas about what it would be like to live with an illness or impairment. Without first- or second-hand acquaintance with the details, their view would probably be based on popular representations of illness and may be limited and anecdotal. This may be particularly acute in cases of illnesses that are rare, medically poorly understood, or otherwise stigmatized. Healthy people would think of the illness or impairment as the defining feature of such a life, and therefore over-weight it—a phenomenon termed 'focusing illusion' by Schkade and Kahneman (1998).

---

[6] I thank one of the anonymous reviewers for this example.

Healthy 'outsider' knowledge of the impact a particular condition may have on their own lives would be limited, and hence their ability to estimate this impact or imagine ways of coping with it would be markedly diminished. Healthy people also fail to anticipate how ill people adapt to poor health (Riis et al. 2005) and therefore fail to consider that such an adaptive process may enable them to have good levels of well-being despite poor health. We can further suggest that some healthy people not only fail to anticipate such adaption but more fundamentally fail to imagine that ill people adapt at all, by focusing solely on the loss brought about by illness.[7]

The sentiment expressed by my taxi driver and the people described by Harriet Johnson is a third-person 'outsider' view. Outsiders do not have access to the experience of illness or disability, but only to their *imagined* experience. As psychologists Daniel Gilbert (2006) and Jonathan Haidt (2006) argue, our ability to imagine counterfactual situations is limited and hampered by general psychological deficiencies. The deficiencies are: limited imagination (the ability to imagine what a counterfactual situation would be like) and flawed memory (defects in recall and recognition), leading to a limited ability to remember what a period of sickness was like and inability to recognize past experiences of resilience and coping in difficult circumstances. In addition, in general people are unrealistically optimistic and tend to overestimate the possibility of good things happening to them in the future (Weinstein 1980). Healthy people spend less time imagining themselves old and unwell or diagnosed with a serious illness than they do imagining themselves on a post-retirement luxury cruise.

Let us look more closely at each of these factors. As Gilbert claims, imagination fails us in several ways. First, there is the focusing illusion mentioned above, about which Gilbert writes:

[...] when sighted people imagine being blind, they seem to forget that blindness is not a full-time job. Blind people can't see, but they do most of the things that sighted people do [...] and thus they are just as happy as sighted people are [...] when sighted people imagine being blind, they fail to imagine all the other things that such a life might be about, hence they mispredict how satisfying such a life can be (2006, p. 104).

---

[7] I thank Ian James Kidd for this last point.

We have little capacity to imagine a complex situation remote from our own life because crucially, we do not have the required detail. Without having first- or at least second-hand knowledge of the complexity of ensuring stable blood sugar levels, the frequent need to draw one's own blood, the calculation of insulin amounts needed prior to having a snack, and so on, it is impossible to imagine what a life with diabetes might be like. I moved from a hazy and abstract understanding of diabetes to a startling recognition of my ignorance when I spent a day with a diabetic friend. The reality of the disease was very different and much more complex than my poorly imagined picture of it.

In addition, our counterfactual imagination suffers from a modal bias: people tend to devise counterfactual scenarios by minor modification of actual states of affairs, not radical changes. They implicitly assess the plausibility of a counterfactual alternative in terms of its modal distance from the actual state of affairs (Gilbert 2006). Thus our ability to imagine a radically different life, in which bodily conditions are substantially different, is limited.

We also have a limited ability to remember past events and feelings. In contrast to our sense of confidence about our memories, it transpires that we are much less able to remember what a situation was like, what an experience felt like at the time, or what we thought and felt in the past. As Gilbert writes:

The elaborate tapestry of our experience is not stored in memory [. . .] Rather, it is compressed for storage by first being reduced to a few critical threads [. . .] or a small set of key features. Later, when we want to remember our experience, our brains quickly reweave the tapestry by fabricating—not actually retrieving—the bulk of the information that we experience as memory (2006, p. 79).

Our memories do not preserve wholly intact 'clips' of past events. Memory preserves core features of the situation rather than the episode in its entirety. Important information is discarded and then reconstructed again. Remembering largely involves confabulating. Hence our ability to remember what it was like when we were sick is compromised. In addition, when such memories are confabulated, the process is likely to be informed by prejudiced expectations, biases, and stereotypes rather than actual facts. Finally, people are more reluctant to remember experiences of illness, and memory is weakest when we fail to exercise it regularly. We have a natural tendency to recall triumphs frequently but to forget defeats, so it is unlikely that we regularly replay a period of illness.

The practical reason human memory has evolved this way is efficiency: by storing only core features of a situation we make considerable savings on memory storage space. This is probably a good enough mechanism for most events. It guarantees personal identity, ensures that salient information is remembered, and that important events are remembered well (even if not accurately). But factual and emotional details of non-salient events are less important in an evolutionary sense. They are therefore forgotten, and when recalled, confabulated. Pezdek et al. (2009) showed participants a short video of a crime. Participants are then asked to recall information that was not present in the video. Many participants provided answers to such questions, showing that confabulation, rather than retrieval, was in operation.

Gilbert summarizes this by saying: 'memories—especially memories of experiences—are notoriously unreliable' (2006, p. 40). Over-confidence with respect to memory has also been shown to be particularly misleading in situations of danger or high emotions. High adrenalin levels or feelings of panic interfere with cognitive function and make memories of an unusual or negative situation even less trustworthy. For example, eyewitness misidentification is argued by *The Innocence Project* to be rife and 'the greatest contributing factor to wrongful convictions proven by DNA testing, playing a role in more than 70% of convictions overturned through DNA' in the US (compare Cutler and Penrod 1995).[8] Similarly, particular states, such as intense physical activity, can reduce recall and recognition ability (Hope et al. 2012).

So we have three reasons to distrust the intuitions healthy people have about illness: they lack the relevant information, they cannot accurately imagine life with illness, and have only rudimentary memories of their own periods of illness.[9] In addition, our implicit views of illness are likely to be shaped by pathophobic prejudices: in a state of informational inadequacy, imaginative limitation, and insufficient memory, a person's

---

[8] See <http://www.innocenceproject.org/causes-wrongful-conviction/eyewitness-misiden tification> (accessed 26 October 2015).

[9] This poses serious questions for frameworks such as 'advance directives', in which people are asked to make a decision about their future medical treatment, should they become unable to speak or lose cognitive capacity (e.g. in dementia). The healthy you, now, cannot make good predictions about what the ill you in the future may want.

intuitions are not just insufficiently grounded, but also much more vulnerable to prejudices arising from negative stereotypes about illness and disability (see Chapter 8).

6.2 Resilience in Illness

Adverse life events such as serious illness, accidents, divorce, or loss of a loved one are often viewed as life-destroying. Illness is no exception to the general finding, replicated in many studies and with respect to different kinds of events: we erroneously think that negative events will affect us more intensely and for longer than they actually do (Haidt 2006, p. 136ff.; Lyubomirsky 2007; Gilbert 2006, p. 152). As Haidt notes, people systematically underestimate their ability to cope with adverse circumstances and have no way of predicting the personal growth, resilience, and development that often follow a period of great difficulty. In fact, he goes so far as to present an 'adversity hypothesis': '[. . .] people need adversity, setbacks, and perhaps even trauma to reach the highest levels of strength, fulfilment, and personal development' (2006, p. 136).

Haidt calls this consequence of experiencing adversity 'post-traumatic growth' and notes three mechanisms that enable such personal growth in the face of adversity: adversity reveals hidden abilities; it 'makes good relationships better'; and it changes priorities in a way that provides focus and peace of mind. I will describe each in turn.

First, confronting a challenge, in the form of an accident, divorce, losing one's job, or an illness, reveals hidden abilities which change one's self-image. 'One of the most common lessons people draw from bereavement or trauma is that they are much stronger than they realized, and this new appreciation of their strength then gives them confidence to face future challenges' (Haidt 2006, p. 139). This is not just a form of self-deception; studies show that people who have suffered hardship recover more quickly when faced with future adversity; they have genuinely become more resilient (2006, p. 139).

The 'adversity hypothesis' tallies with the view of suffering as morally or spiritually rewarding and edifying (Kidd 2012). In these cases the suffering can be mitigated by a deep sense that one is suffering for a purpose. Adverse experiences are often described as a gift, trial, journey, or test, of which possibly the most famous one is the story of Job, whose faith was tested by God through a series of calamities. This spiritual growth as a positive outcome of adversity may occur whether or not the suffering actually was purposeful, since the very belief that it is could be positively motivating. For example, the pain of childbirth is often described as being mitigated or interpreted very differently from other kinds of pain because of its productive and life-bestowing purpose (Heyes 2012).

Not all trauma is followed by 'post-traumatic growth'. Indeed, trauma can be permanently life-shattering, destroying one's sense of trust in the world and in others (Ratcliffe et al. 2013). We do not need to—and indeed should not—agree with the strong claim that *all* trauma, regardless of its severity, results in growth. Traumatic events in early years can destroy trust in ways that permanently maims the child's future psychology and behaviour (Getz et al. 2011; Gerhardt 2004). Pre-birth trauma such as exposure to alcohol, domestic violence, or high cortisol levels *in utero* can permanently damage executive function, memory, and impulse control. Trust can also be irrevocably destroyed by trauma in adulthood, which can be permanently disabling, as can be seen in cases of post-traumatic stress disorder (PTSD). I suggest that some events that may initially be experienced as traumatic or difficult (such as a divorce) may result in post-traumatic growth, in ways that are unexpected even to the person undergoing the adversity herself.

The second factor promoting post-traumatic growth is improvement to relationships noted in circumstances in which people are forced to speak frankly about important issues, such as death and incapacitation, and are forced to ask for help. The lack of intimacy possible in routine social encounters becomes untenable and an opportunity emerges for authentic relationships to become stronger and for honesty and intimacy to be sought.

Bronnie Ware (2012), a palliative care nurse, writes about caring for a dying woman: 'once we reached this level of honesty, our conversations flowed unhindered. There was no time for holding back [. . .]. With death on her doorstep, Elisabeth, too, enjoyed the openness of our constant exchanges' (2012, p. 149). The actor Christopher Reeve, who became paralyzed from the neck down in a sports accidents, said that he did not appreciate other people nearly as much before he became severely disabled (quoted in Gilbert 2006). When adversity strikes, Haidt notes, it 'strengthens relationships and opens people's hearts to one another' (2006, p. 139).

The third factor promoting post-traumatic growth is the change to priorities and values, which makes people who undergo adversity focus on the present. This has been noted in qualitative studies of people who have become ill (see for example Brennan 2012; Frank 2010). The unpredictability of the course of illness leads some to adopt a perspective that focuses on living in the present and refrains from looking towards the future, making long-term plans, or having rigid goals (Michael 1996; Carel 2013a). Such focus on the present leads to greater enjoyment of, and attentiveness to, one's current experience. In the case of illness in particular, because of physical and mental limitations, pain, fatigue, and sometimes a poor prognosis, the emphasis on what is still possible is described by some ill people as a positive way of dealing with the effects of illness. This is not something that happens automatically; it is the result of a process of adjustment and reflection.

The challenge in the case of illness, and in particular when prognosis is poor, is clear. Resilience needs to be developed in response to a considerable challenge. If we return to the phenomenological concepts discussed in the previous chapters, we can think about illness as modifying the ill person's way of being. It can do so in one of two ways. Illness can inflict suffering and limitations that force the ill person to modify their world (e.g. reduced mobility often leads to a shrinking world). This process can be unreflective and reactive, with the ill person simply making the necessary practical adjustments, or giving up prior habits without any reflection on the meaning of such change. Lived unreflectively, such a process can lead to anger and frustration, and these too can remain pre-reflective or entirely unarticulated to the ill person herself.

However, pre-empting Chapter 9, in which illness is viewed, *inter alia*, as an invitation to philosophize, illness can also be seen as dramatically changing the ways of being that are available to a person and thus prompting, provoking, or inviting them to modify their being. This process is one of reflective shaping and guiding of one's way of being. In this case, too, illness forces the ill person to modify her way of being, but she must play an active role in that modification, by reflecting on and guiding the change.

A localized response will not suffice. Illness changes everything about a person's life—their relationship to themselves, their loved ones, their environment, other social relations, their experience of their body, their goals and plans, and their view of the future (see Chapter 3 and

Carel 2013a). Moreover, the challenge of illness is often to learn to live well with facts that we would dearly wish to be otherwise. As I wrote in my book, *Illness*:

> The future no longer contains the vague promise of many more decades. Death is no longer an abstract, remote notion. The soft-focus lens is replaced by a sharp magnifying glass through which terminal stages of illness can be viewed in nauseating detail. The future curls in on itself and at once becomes both exposed and radically curtailed (2013a, p. 143).

The response to this challenge is to turn from taking time for granted to cherishing it. The result is awareness of the present moment and an increased ability to experience fully what seem like mundane moments:

> Time did change for me. I began to take it much more seriously. I began to make a point of enjoying things thoroughly: memorizing sensations, views, moments. Partly in preparation for days to come in which I may not be able to leave the house or my bed, but also in order to feel that I have taken the time to really sense, really experience pleasurable things. I wanted to feel that I am living life to the full in the present. That I *am* now (Carel 2013a, p. 144).

When the future becomes uncertain or just plain bad, it releases us from its usual grip. We normally worry about the future, plan, strategize, strive towards goals. When this element is weakened, many ill people find that turning to the present enables them to slow down, appreciate the moment, and cease worrying about what may happen. The fragility and preciousness of the present become more visible and more appreciated, and this can lead to flourishing in unexpected ways. As Nietzsche (2004) notes about his lengthy period of illness, physical frailty has a regenerative effect:

> It was as if I discovered life anew, myself included; I tasted all the good things, even the small ones, as no other could easily taste them—I turned my will to health, to life, into my philosophy [...] the years when my vitality was at its lowest were when I stopped being a pessimist (2004, p. 8).

Other thinkers, like Epicurus, argue that living in the present is key to well-being. For Epicurus, to achieve well-being is to remove all sources of anguish and pain. Once this state has been attained, nothing more can be added to one's well-being. One has already achieved tranquillity (*ataraxia*), and painlessness (*aponia*) (Epicurus 1984). But in order to achieve tranquillity, an important temporal focus must be achieved first. We must remove our unhealthy connections to the past (regret, longing, bitterness, loss) and the future (anxiety, obsession, worry, desire). Such detachment from the past and the future brings with it an ability to focus on the present—what is happening now, what is being experienced now. Learning to free oneself from the past and the future in order to fully concentrate on the present is an important part of well-being (Carel 2009).

Epicurus recommends that this practice be followed by an effort to bring pleasure and joy into the present, and finally, realizing that the quality of the present moment does not depend on how long it lasts, or how many desires it fulfils (Hadot 1995). A moment can be simple, quaint, unremarkable but for its joy. Adversity, in its many shapes and forms, can be a powerful reminder of the value of such a moment. As Epicurus reprimands us, 'you are not in control of tomorrow and yet you delay your opportunity to rejoice' (1994, p. 36).

This focus on the present is particularly important in the case of illness. Privileging the present can make a significant change to the experience of illness. It enables us to locate the source of mental anguish in the past: memories of things she could once do may cause an ill person great suffering. Similarly, as we saw above, the future may cause great anxiety: one may fear future suffering and death. Focusing on the present can help us distance ourselves from both past and future, and this can be a way to distil a moment of well-being. Focusing on present abilities, joys, and experiences instead of worrying about a no-longer-existing past (my previous healthy self, regrets about its loss and longing for it) and a not-yet existing future (the imagined future suffering that may or may not happen) is a way of avoiding some of the suffering caused by illness.[10]

Qualitative studies of the experience of a variety of illnesses have identified central positive themes that emerge from coping with illness.

---

[10] Gilbert offers another psychological mechanism that explains resilience in illness, namely, rationalization. People consider their situation relative to that of others in the relevant reference group, in an attempt to minimize the perceived adversity of their situation by contextualizing it and comparing it to those in a similar predicament. For example, in a study by Taylor et al. (1986) 96 per cent of cancer patients claimed to be in better health than the average cancer patient. People with life-threatening illnesses are likely to compare themselves with those who are in worse shape and this protects them from thinking that their own state is substantially worse than that of the healthy. However, this mechanism has a major fault: it is a form of self-deception.

For example, themes such as being courageous, regaining control over an altered life course, reshaping the self, self-transcendence, empowerment, and discovery, are often described by interviewees suffering from an illness (Thorne and Paterson 1998). These themes seem to support the adversity hypothesis, as they express subjective articulations of post-traumatic growth.

As Helson (1964) notes, well-being is comprised of multiple components, and it is possible that a reduction in one area (health) is compensated for by an increase in other areas (personal growth, intimacy, relationships), so the overall level of well-being remains the same. Helson's observation ties in with a general view of illness as a non-linear, changeable experience. Paterson (2001) developed a 'shifting perspectives' model of chronic illness, in which illness comes into the foreground during the initial phase of illness, but then recedes into the background, coming back into the foreground during episodes of symptom exacerbation or disease progression. This model explains how illness may fade into the background and accounts for periods in which it plays no substantial role.

A final way of explaining resilience in illness relates to how we experience our body. As discussed in Chapter 2, some phenomenologists have suggested that the body in health is transparent (Sartre 2003), hidden (Keane 2014), or even absent (Leder 1990) and this has often been contrasted with illness, in which the body becomes cumbersome and aberrant, and is thematized mainly through negative medical attention (Toombs 1987). As I suggested in Chapter 2, although there is a discontinuity in bodily experience, this does not mean that the healthy body is impervious to failure, pain, and damage. Although the change in bodily experience when one becomes ill is pronounced, indeed dramatic, the healthy body contains its own opaqueness.

I suggest that glimpses of the experience of illness, although the experience itself is radically disruptive, can be seen in everyday bodily failure. If we accept that the body is vulnerable even when it is generally healthy, we may also be more open to the suggestion that illness often undergoes a normalizing effect over time. This is not to say that illness is not a dramatic break from the habits of health, as I argued in Chapter 2. But adaptive responses to illness can be taken to demonstrate the plasticity and adaptability that characterize human behaviour more generally and enable us to get used to radically different (and radically curtailed) forms of embodiment. This adaptability makes illness more tolerable and those who are ill more resilient.

This sits well with the ideas discussed above, such as hedonic adaptation, the focusing illusion, and the empirical finding that ill health does not impact on well-being over time, beyond an initial period of disruption. Bodily experiences which initially seem bizarre and extraordinary become quotidian once they are added to our bodily repertoire. The same applies for loss of abilities and possibilities. Perhaps we similarly adjust to such losses by shrinking our perceived horizon of possibilities.

To illustrate this point, here are two first-person accounts of such a process. The first is a multiple sclerosis patient, reflecting on the fact that he can no longer walk or even stand. These abilities, he says, lost much of their importance and are 'no longer within the sphere of possibility and are therefore not missed as though they were possible' (Schneider 1998, p. 71). The second example comes from my account of my own illness:

My body adapted with astonishing alacrity to new limitations. My mind quickly forgot how things were before. Within a year my physical habits were entirely different. Whereas in the first months my body would attempt a brisk pace, hurrying up stairs, physical impatience, these movements have been erased from my bodily repertoire. While my memory still contained images of mountain views and the inside of a gym, I could no longer remember what it felt like to run, to work out, the euphoric sensation of healthy exertion, the effortlessness of being young and healthy. New habits were formed and a new way of negotiating the world was incorporated into my physicality. Blissful forgetfulness of the pleasures of physical movement accompanied them (Carel 2013a, p. 36).

As was argued in Chapter 2, it is not clear that the healthy body is as transparent or as absent as has been postulated in the phenomenological literature. The body continuously emerges for us in everyday activities. It may not be the thematic focus of our activity or attention, but it will still provide constant reminders of itself through hunger, fatigue, failure to concentrate, or an inability to learn a new task. The body is constantly there for us not merely as a peripheral enabling background, but as the medium of our experience, which necessarily includes a self-reflexive dimension, although this can recede into the background. When we experience the beauty of an icy winter landscape, feeling cold is part of this experience.

The body is not transparent in health—and nor is it entirely opaque or obstructive in illness, as discussed in Chapter 3. Bodily enjoyment is often still possible, at least in some form, and the increased opacity that characterizes illness may be mitigated by other factors. Several authors in the phenomenology of illness have emphasized the opacity and limitations of illness (Toombs 1987; Svenaeus 2001). But it is also important to acknowledge the possibility of continued, or regained, normalcy, and the periods of stability that often characterize chronic illness. This has been captured in the notions of 'health within illness' (Lindsey 1996) and 'well-being within illness' (Carel 2007), which aim to make space for the possibility of being well and happy within the broader context of having a chronic condition or impairment.

Even if the body in illness becomes more opaque and cumbersome, this opacity has the same changeability and varied inflections as the putative transparency of health. I do not wish to belittle the effects of ill health, the suffering, loss of freedom, or indeed shock and shattering of hopes and plans. I do want to point out that these negative effects may be transient and may only temporarily dominate experience, later receding to the background. They can be mitigated by life's other goods, provided that they are there, such as intimacy, finding joy and solace in new activities, and the ability to adapt to what might seem at the outset as an intolerably cruel turn of fate.

6.3 Well-being as Achievement

This chapter posed a question, namely, is well-being possible within the confines of poor health, an uncertain prognosis, and limitation of one's freedom? I suggested that two enigmas of health, to borrow Gadamer's phrase, arise from this question. First, why don't ill people become markedly unhappy as a result of their illness? Second, given that we have robust evidence showing that well-being is not much reduced by ill health, why do we fear illness and see it as a terrible evil when we are healthy?

I suggested some explanations for the remarkable resilience people show in the face of illness (and some other trauma), including the adversity hypothesis, and evidence from psychology, qualitative health research, and first-person accounts of illness. I sought to explain the discrepancy by looking at differences between 'insider' and 'outsider' views of illness and disability.

I would now like to look more closely at the notion of well-being. Many definitions of well-being, flourishing, and happiness have been proposed by philosophers (for a selection of philosophical writings on happiness see Cahn and Vitrano 2008). We find Aristotle defining flourishing objectively, as *eudaimonia*, a life that consists of cultivating virtue, where virtue is objective. We find psychologists turning to subjective measurements of well-being as definitive, and arguing that 'no one but you knows or should tell you how happy you truly are' (Lyubomirsky 2007, p. 34). We find behavioural economists developing sophisticated methods of showing that sometimes even the subject herself is subject to ignorance, bias, and illusion when she tries to assess her own well-being (Schkade and Kahneman 1998). And we find that even our own attempts to predict what will make us happy are hampered by the limits and fallibility of our memory and imagination (Gilbert 2006).

This does not leave us in a very comfortable, or consistent, position with respect to the notion of happiness. The weighty issue of what happiness is, how to define it, how to ensure that participants in empirical studies understand it, whether it is a real first-order phenomenon or a construction, and whether it is too subjective to be meaningful, and other such questions, deserve full philosophical treatment that is also well informed about the empirical literature in other disciplines. Excellent and distinctively philosophical treatments of these questions have been provided by Daniel Haybron (2010), Valerie Tiberius (2010), and Fred Feldman (2010), among others. The question of well-being in relation to health has also been comprehensively discussed by Dan Hausman in his recent book, *Valuing Health* (2014).

What I would like to suggest here is that we put to one side these conflicting definitions and methodologies and turn, in phenomenological spirit, to what ill and disabled people themselves say about well-being within the constraints of illness. Here is what disability activist Harriet Johnson (2003) wrote:

Are we 'worse off'? I don't think so. For those of us with congenital conditions, disability shapes all we are. Those disabled later in life adapt. We take constraints that no one would choose and build rich and satisfying lives within them.[11]

And here is what I wrote about illness and the sense of helplessness that characterizes being ill:

I cannot change reality; my illness is here to stay. But I can control some of the elements making up my life. I can, for example, control my thoughts (to some

---

[11]   See <http://www.nytimes.com/2003/02/16/magazine/unspeakable-conversations.html? pagewanted=all> (accessed 19 October 2015).

extent); I can control my reactions; I can cultivate the happy aspects of my life and I can say no to distressing thoughts and actions. I can choose what to do with the time I have and I can reject thoughts that cause me agony. I can learn to think clearly about my life, give meaning even to events beyond my control and modify my concepts of happiness, death, illness and time (Carel 2013a, p. 150).

What these passages seem to suggest is that well-being in the context of illness is neither something that is impossible, nor something that can be taken for granted. Rather, cultivating well-being within illness and learning to live well with physical and mental constraints requires conscious effort and is an *achievement* that should be recognized as such. Julia Annas suggests that happiness is 'the task of forming my life as a whole in and by the way I act' (2008, p. 242).

If we take seriously the phenomenological approach to illness and the robust evidence that ill and disabled people are no less happy than other people, we can conclude that paying close attention to first-person accounts may yield important insights about the experience of illness. If happiness is an achievement that requires thought, planning, and work, this view contributes to our understanding of why illness does not affect long-term well-being. Illness provides us with a context and opportunity for the kind of reflection and revaluation that are the condition for happiness on Annas's account. So it is no wonder that ill people find ways of being happy, even within the constraints of illness.

7

Illness as Being-towards-Death

We ended the previous chapter with a discussion of living in the present as a mode of well-being that can be experienced within illness. Such living in the present requires bracketing worries about the future. Even if we train ourselves to focus our attention on the present, and are largely successful in that, we are still temporal beings with links to both past and future. These links structure our way of being in the present.

Both temporal modes, past and future, play an important role in illness. The past, as discussed in the previous chapter, can be a source of regret and grief, because it contrasts the current state of illness with the past state of health. The future can be a source of anxiety about future treatment, disease progression, and ultimately, especially in the case of a poor prognosis, death. Death is positioned on the existential horizon of illness, and therefore makes up part of the illness experience. But death also has a presence in health, because life and death are intimately linked. Death structures life, limits it, and shapes our goals, choices, and actions (Carel 2006). It is this mutual implication of life and death that is developed in this chapter.

With the exception of suicide and accidents, there is no death without prior illness, even if it is as brief as a stroke or cardiac arrest. Yet the philosophical literature on death rarely discusses the importance of illness within this context. This chapter argues that illness has a close relationship to both dying and death, and that severing that link has been detrimental to the philosophical understanding of death and finitude. Indeed, illness is a 'being towards death' in its most intense form. I propose using Heidegger's notion of 'inability to be', discussed in Chapter 3, §4, to make this connection. If illness is characterized by a degree of 'being unable to be' death can be seen as total inability to be, the closure of all existential possibilities.

There are other significant links between illness and death. In previous chapters we considered how illness impacts on the ill person's life and changes her way of being in the world, her embodied comportment in the physical and social environment, and her values and goals. However, serious illness has another distinctive feature: it sometimes leads to the end of lived experience and closure of one's experiential horizons. In other words, serious illness is often the first sign of the end of life, and hence reveals that human life is finite. This finitude, I suggest, has radical consequences for how we live, which are explored in this chapter. Illness marks the transition from existence to non-existence, and in that sense is intimately related to, and a herald of, our finitude.

But what is there to say about death from a phenomenological perspective? If death is the complete absence of all experience, an experiential blank, what can phenomenology—the study of human experience—say about it? We can answer in one of two ways. First, we can distinguish *death* from *dying*. A phenomenology of dying is entirely possible; indeed, the experience of dying is an important theme of a phenomenology of illness. Second, we can phenomenologically examine the relationship we have with death as living finite creatures who have an understanding of their mortality. While the earlier chapters of this book offered the former, namely, a phenomenology of serious illness, this chapter offers the latter: a phenomenology of existing as finite, of being towards death.

Here, I use Heidegger's account to suggest that death is intrinsic to life and thus relevant to both healthy and ill existence. I claim that death and finitude are structuring components within existence, breaking with the view of life and death as mutually exclusive (Carel 2006). Illness is a bridge between the two and as such plays an important, yet largely unacknowledged metaphysical and epistemic role in human life.[1] Heidegger's formulation of existence as being-towards-death (*Sein-zum-Tode*) conceptualizes life and death, possibility and limitation, as intimately linked.

Death influences life as a limit that must constantly be taken into account, even within the context of everyday activities and projects. If death shapes how we live and think about temporality, possibility, action, and decision, existence must be understood as structured by death.

---

[1] The epistemic role referred to here is one of edification (see Kidd 2012). The next chapters discuss these themes.

This is not only an *epistemic* claim about the influence an awareness of our mortality has on how we live, but an *ontological* claim about the structure of human life. Heidegger calls this structure Dasein, and his analysis of Dasein as finite temporality is the focus of this chapter.[2]

Although the issue of finitude is expressed here as a philosophical matter, it is obvious that our attitudes towards death are not purely theoretical. Death is a highly personal problem we each have to confront. Death always belongs to a specific Dasein: it 'lays claim to it as an individual Dasein' (Heidegger 1962, p. 308). Although someone could sacrifice their life for someone else and 'die for them', that would only postpone death, because ultimately 'no one can take the Other's dying away from him' (1962, p. 284).

However, our attitudes towards death are also largely shaped by society and culture.[3] These attitudes attempt to deny death by cultivating an attitude of indifference towards it or by bracketing it as irrelevant to life. An excellent example of such bracketing comes from Epicurus (1994), who states:

So death, the most frightening of all bad things, is nothing to us; since when we exist, death is not yet present and when death is present, then we do not exist. Therefore, it is relevant neither to the living nor to the dead, since it does not affect the former and the latter do not exist (1994, p. 29).

Such 'tranquilization', claims Heidegger, encourages a 'constant *fleeing in the face of death*' (1962, p. 298). These kinds of attitude to death alienate Dasein from its finite existence by discouraging an existential understanding of it and therefore make Dasein inauthentic.[4] Accordingly, there is a tension between the personal demand on each of us to think about our own death and the public attitude to death, which is often *thanatophobic*, as can be seen in the modern removal of the process

---

[2] Throughout this chapter I will use Heidegger's term, Dasein, to denote a human being.

[3] For a classic historical account of changing attitudes towards death see Jacques Choron (1963, 1972).

[4] The German term for authenticity, *der Eigentlichkeit*, has been interpreted, analysed, and criticized extensively since 1927, particularly in existentialist philosophy and Theodore Adorno's critical work, *The Jargon of Authenticity* (1973). The double meaning of both 'proper' and 'property' in the German '*Eigen*' is retained in alternative translations of the term as 'appropriateness' (Stambaugh) or 'ownedness' (Carman), but the emphasis on the relationship to the self, which resonates in 'authenticity', tends to be lost. I therefore retain the Maquarrie and Robinson translation of *der Eigentlichkeit* as authenticity, while bearing in mind the inflated resonance of the English term.

of dying and of the dead from the community into facilities such as hospitals and morgues, as well as the accompanying disposal practices.

When considering death, the tension between the individual and society becomes pertinent: whereas *das Man* (the public attitude, according to Heidegger, translated as the 'They', 'the One', or 'the Anyone') aims at tranquillization, Dasein must free itself from this tranquillization in order to face death authentically.[5] Thinking about death requires taking into account both its personal and public dimensions. This allows the meaningfulness of death to emerge as a comprehensive structure, reflecting Dasein's inherently social nature. This dual nature also applies to illness. Illness has a strong first-personal and private nature, but it also has important social consequences, for example for others' expectations, social arrangements, and the apportioning of responsibility, as documented by Talcott Parsons in his discussion of the 'sick role' (1971) and Erving Goffman in his discussion of stigma (1963).

For Heidegger, there is a deep difference between my death and that of another person, claiming that from the first-person perspective death is the radical closure of all possibilities of existence, whereas from the second- and third-person perspective, death is an event in life.[6] This led him to claim that Dasein can encounter its own death authentically, but not the death of another. Heidegger does not discuss the possibility of an authentic experience of the death of another. This lacuna creates what is in my view an unjustified identification of individuation and authenticity, and can be overcome, as I propose below.

There are good reasons to weaken the strong opposition Heidegger posits between my death and that of another, in order to make room for the possibility of a shared authentic experience of illness and of dying.

---

[5] The German term *das Man* refers to a generic other, and is used in the same way as the English 'one'. It is used to express views or actions without attributing them to any particular individual. 'One would think that the president ought to resign', or 'they say it is going to rain' are examples of such impersonal attributions. Heidegger uses *das Man* to indicate an internal agency made up of social norms and conventions that eclipses the authentic self. Heidegger defines *das Man* as an *existentiale*, part of *Dasein*'s ontological structure. The English translations of the term are 'the "they"' (Macquarrie and Robinson, Stambaugh), 'the One' (Dreyfus, Carman), or 'the Anyone' (Kisiel). I use the German, which has by now become widely familiar.

[6] It remains open what Heidegger's account might make of a second-person perspective on death. I suggest that witnessing the death of a loved one is a profound existential change that should be positioned between the first-person complete annihilation and the third-person perspective, for whom death is merely an event in one's life.

I suggest that the rift Heidegger introduces between my death and the death of another rests on an understanding of authenticity and inauthenticity as mutually exclusive. Rejecting this understanding enables us to view being-towards-death as relational, and rethink the individuation requirement as the condition of authenticity.

It is important to note at the outset that Heidegger's concept of death is different from the ordinary concept. As Taylor Carman points out, 'An enormous amount of confusion has resulted from the fact that by "death" Heidegger does not mean quite what is commonly meant by the word. But neither is his existential conception of death wholly alien to our ordinary understanding' (2003, p. 276). By 'death' (*Der Tod*) Heidegger does not mean what we commonly understand as death, namely, the passing away or termination of an individual's life. This point has been overlooked in the literature on Heidegger's death analysis and has underpinned criticisms levelled at Heidegger's concept of death (Levinas 1969, 1998; Edwards 1975, 1976, 1979; Philipse 1998; Chanter 2001, Sartre 2003).

The first two sections of the chapter present Heidegger's notion of death and address two of its key features: the distinction between death and demise, and the precise meaning of the term 'possibility' (*die Möglichkeit*) in *Being and Time*. They show how by reinterpreting Heidegger's concept of death we can achieve a fuller, coherent understanding of the concept. This is achieved by presenting two interpretations of Heidegger's concept of death offered by William Blattner and Hubert Dreyfus. The next section argues that there is a missing dimension from their interpretation, namely the dimension of finitude, and suggests a way of augmenting the interpretation. The chapter ends by suggesting that a relational understanding of death is possible within the Heideggerian framework, and moreover would be a positive step towards ameliorating the loneliness of illness and dying.

7.1 Death Structuring Life

Death plays a central role in Heidegger's early work, in particular *Being and Time*, and is a crucial element of Dasein's structure.[7] In the 1929–30

[7] This chapter focuses on Heidegger's early treatment of death, mainly in *Being and Time* (1927), but also in the lecture courses from that period: *History of the Concept of Time*

lecture course *The Fundamental Concepts of Metaphysics* Heidegger writes: 'Finitude is not some property that is merely attached to us, but is *our fundamental way of being*' (1995, p. 5). Death defines and shapes Dasein's existence as its limit. It is 'the limit-situation that defines the limits of Dasein's ability-to-be' (Blattner 1994, p. 67). This limit becomes existentially significant because of Dasein's unique capacity to anticipate it, a capacity that structures everyday life as 'being-towards-death'.

But Dasein does not experience its own death. Death is not a phenomenon contained within Dasein's experiential horizon. Heidegger is not offering an analysis of what he calls demise (*ableben*), the event that ends Dasein's life, transforming a living body (*Leib*) into a corpse (*Körper*). Rather, his analysis focuses on how Dasein's existence is shaped by mortality and how life is a process of dying (*sterben*). Heidegger does not offer a phenomenology of death, but an analysis of being-towards-death, a phenomenology of finite existence (1962, p. 277). This led Françoise Dastur (1996) to say, 'There is a phenomenology not of death, but of our relatedness to death, our mortality' (1996, p. 42). Therefore, she claims, the analysis focuses not on death but on being in a relation to death.

Equally, Heidegger does not present us with a phenomenology of the death of others. Although someone else's death could be a profound event for me, I am barred from experiencing their death. At most, I experience my loss. As Heidegger writes, 'Death does indeed reveal itself as a loss, but a loss such as is experienced by those who remain. In suffering this loss, however, we have no way of access to the loss-of-Being as such which the dying man "suffers"' (1962, p. 282).

Because of death's inaccessibility it can only be related to by scrutinizing its effect on life. 'Phenomenologically speaking, then, life is death's representative, the proxy through which death's resistance to Dasein's grasp is at once acknowledged and overcome' (Mulhall 2005, p. 305). Heidegger does not focus on the moment of demise, which is

((1992) lectures given in 1925), *The Basic Problems of Phenomenology* (given in 1927), *The Metaphysical Foundations of Logic* (given in 1928), and *The Fundamental Concepts of Metaphysics* (given in 1929–30), as well as *Kant and the Problem of Metaphysics* ((1990) lectures given in 1929). The later Heidegger attributes importance to death, as seen in the use of the term 'mortals' (*die Sterblichen*) replacing '*Dasein*', but there is no sustained treatment of death in the later work. I therefore centre this study of Heidegger's treatment of death and finitude on the early period, focusing mainly on *Being and Time*.

phenomenologically inaccessible, but on the anticipation or forerunning (*vorlaufen*) towards death that constitutes Dasein as mortal. Such forerunning can be a constitutive element of serious illness and in ageing. We can thus say that illness provides a view of mortality that can profoundly change one's self-understanding.

The phenomenological investigation of death must therefore look to life and everyday existence, examining how they are shaped and affected by death. Although death is not an event within life, death influences everyday existence and Dasein's structure, which Heidegger characterizes as 'being-towards-death'. In a passage from *History of the Concept of Time* Heidegger writes:

> The certainty that 'I myself am in that I will die' is *the basic certainty of Dasein itself* [. . .]. If such pointed formulations mean anything at all, then the appropriate statement pertaining to Dasein in its being would have to be *sum - moribundus*, *moribundus* not as someone gravely ill or wounded, but insofar as I am, I am moribundus. *The moribundus first gives the sum its sense* (1992, p. 317).

Death constitutes Dasein as temporally finite: 'Only in dying can I to some extent say absolutely "I am"', writes Heidegger (1992, p. 318). This changes Dasein's temporal structure, which is now stated to be finite. It also changes one's projection (*entwerfen*), discussed in Chapter 3 and now revisited in light of Dasein's finitude. Dasein constantly projects itself towards its future by making plans and choosing to pursue certain possibilities. In this projection Dasein also projects itself towards death, its impossibility to be or to have any more possibilities. Death is the possibility of no longer being able to be (1962, p. 294).

Dasein's movement towards the future is a movement towards annihilation. Dasein presses into the future by projecting itself towards its chosen possibilities. But this movement is bound by Dasein's past choices and actions, as well as by some of its given features. Dasein is historically and socially situated in ways that were not chosen by it, as denoted by the notion of thrownness (*Geworfenheit*). Dasein is further limited by its finitude of possibility and by death. Dasein is therefore *thrown projection*.

This formulation expresses the fact that the freedom to press into a certain possibility, the freedom to shape one's future, is not unbound. Rather, thrown projection captures the idea of *bound* or *finite* freedom.

The task is therefore 'to conceive freedom in its finitude, and to see that, by proving boundedness, one has neither impaired freedom nor curtailed its essence' (Heidegger 1984, p. 196).

In addition, death is not an ordinary aim that Dasein can project itself towards. While all other possibilities give Dasein something to be, death is the closing down of Dasein's temporal structure. It is also different from other possibilities because it is unavoidable. This necessity generates a difference between my death and the death of others. All events, including the death of others, take place within my world and are therefore subsumed under my experiential horizon. But my death is the closure of my experiential horizon: it is the possibility of the impossibility of existence (Heidegger 1962, p. 307).

Dasein not only has a finite structure, but is also endowed with the ability to conceive of its death: Dasein understands that it is going to die. Death is not only an ontological fact structuring Dasein as temporally finite; this fact is also necessarily reflected in Dasein's life. Therefore death is not only an external endpoint but bears internally on how Dasein lives and what kind of projects and choices are open to it. Finitude shapes the projects and plans we make and is therefore implicit in our self-conception.

Because death is a constant condition of all events within life, the phenomenological project of understanding death is a demand to understand life as finite, rather than focusing on the event that ends Dasein's life. Heidegger therefore introduces a distinction between Dasein's dying (*sterben*) and demise (*ableben*), and the perishing (*verenden*) of animals which lack awareness of their finitude.[8] For Dasein, death is something it must face. It is this *relation* to death that Heidegger explores, in his analysis of being *towards* death.

This projection towards death is problematic: phenomenologically it is a projection towards something that cannot be experienced; ontologically it is a projection towards something that is *not*, towards annihilation. Death destroys Dasein, who constantly moves towards it. Although the

---

[8] This claim is problematic and inessential to Heidegger's view. It is problematic because it reintroduces Cartesian mind/body dualism in the form of the reason (*Dasein*)/nature (animal) divide. It is inessential because some animals may be *Dasein* too and have the capacity to sense their finitude, and this does not undermine Heidegger's argument. Also, awareness of one's mortality comes in degrees (compare Plumwood 1993).

movement is certain, the time and manner of Dasein's death remain indefinite, so it is a constant shadow.[9]

The more unveiledly this possibility gets understood, the more purely does the understanding penetrate into it *as the possibility of the impossibility of any existence at all.* Death, as possibility, gives Dasein nothing to be 'actualized', nothing which Dasein, as actual, could itself *be* (Heidegger 1962, p. 307).

Heidegger characterizes death as *ownmost* (*eigenst*), *non-relational* (*unbezüglich*), and *not to be outstripped* (*unüberholbar*). 'Ownmost' indicates death's essential belonging to Dasein, identifying death as something that cannot be taken away from a particular individual or passed on to someone else (1962, p. 284). Death is different to other attributes or events that can be performed by or given to a different Dasein. Someone else can teach my class for me if I am ill, or donate blood instead of me. But death is my ownmost because even if someone else sacrifices their life for me, I still have to die myself. As Freud comments in 1915: 'everyone owes nature a death and must expect to pay the debt' (1985, p. 77).

The second characteristic, non-relationality, expresses death's individuating effect on Dasein. Death severs Dasein's links to its world and the entities within it. The third characteristic, not to be outstripped, is a combination of two further attributes, death's *certainty* and *indefiniteness*. Because it is certain, the threat of death hangs over us constantly. Because it is indefinite, that is, we do not know when we shall die, we are constantly threatened by it. As a result, we cannot outstrip or overtake death; we cannot hold it fixed. For example, we look forward to a party; it takes place and we can later view it as a past event that has been temporally surpassed. Death cannot be similarly surpassed. As long as Dasein exists, death is always in front of it and always indefinite.

A poignant example is given in Nabokov's 1935 novel *Invitation to a Beheading*. The protagonist, Cincinnatus, is on death row. Cincinnatus is repeatedly tormented by false alarms of his impending execution given by his sadistic gaoler. Cincinnatus says, 'The compensation for a death sentence is knowledge of the exact hour when one is to die. A great luxury, but one that is well earned. However, I am being left in that ignorance which is tolerable only to those living at liberty' (1959, p. 14).

---

[9] This uncertainty disappears in the case of suicide, which perhaps explains the sense of relief experienced by those who decide to end their life once the decision has been made.

Death is something which 'Dasein itself has to take over in every case' but which cannot be controlled, surpassed, or temporally determined (Heidegger 1962, p. 294).

Dasein constantly runs (*vorlaufen*) towards its death. It constantly *anticipates* death; death is always and only something that is yet to come. We never expect its actualization, because death gives us nothing to be actualized (1962, p. 307). Anticipation does not denote expecting or ruminating. Rather, to anticipate death is simply to live as finite and to understand one's temporal structure fully. For Heidegger this understanding of one's finitude reveals Dasein's ownmost ability-to-be (*Seinkönnen*) and the possibility of authentic existence (p. 307). Anticipation is very different to expecting (*erwarten*) an event, because expecting means waiting for an actualization. Death cannot be actualized. It is the completion of Dasein, but paradoxically this completion is annihilation. As long as Dasein exists there is a part of it that is still lacking and towards which it constantly moves (1962, pp. 293–4). Dasein cannot be a totality because its completion is also its annihilation (p. 286).

Death is unlike the completion of a painting, the end of a road, the ripening of a fruit, or finishing a loaf of bread by eating the last slice (1962, p. 289). The comparison to ripening fruit is particularly salient, because its fulfilment or perfecting (*vollenden*) is also its end. Dasein, too, fulfils its course with death, but leaves nothing behind. Death is an end, but not a teleological end. By juxtaposing Dasein's end with other kinds of ending Heidegger reveals the untimeliness of death. Dasein may die before it fulfils its projects, but may well have passed its prime before it dies (p. 288). As a result, for the most part, Dasein ends in an untimely manner, unfulfilled.

Being towards death is an active and practical position. 'Death is a way to be, which Dasein takes over as soon as it is' (p. 288.). Dasein's way of being towards death reveals itself in its choices of possibilities towards which it projects itself, which constitute Dasein's movement towards its future. When Dasein anticipates death it frees itself, because death illuminates all other possibilities as part of a finite structure. Seeing itself as a finite structure enables Dasein to see itself as a whole. This understanding is not theoretical but enacted. Therefore Dasein not only understands itself as a finite whole but *exists* as one.

There are two ways for Dasein to respond to its mortality: authentically and inauthentically. Dasein can choose to respond authentically to death by resolutely anticipating it. This opens the possibility for Dasein to authentically engage with its existence, since it has now grasped it as finite. Dasein can also flee from death by dismissing it as irrelevant to the present. Heidegger calls this attitude 'inauthentic'. The two attitudes to death underlie everyday practical concerns and engagement with the world because our actions are performed within a temporally finite horizon. As a result, no one is exempt from having some sort of attitude towards death. Whether Dasein assumes an authentic attitude towards death or flees from its mortality, it is always bound by death. Death determines Dasein's relationship to its future and its conception of itself as finite and temporal.

7.2 The Meaning of Death

The precise meaning of being-towards-death is not immediately clear, as the secondary literature on Heidegger suggests (Philipse 1998; Mulhall 2005; Edwards 1976; Blattner 1994). Is the formulation of Dasein as being-towards-death intended to capture the insight that life is a constant movement towards death? Or is it the unique capacity of Dasein to understand itself as finite? And if death is simply the absence of existence, then how could it be a possibility? Surely death is the impossibility of all possibilities, and moreover not only possible, but necessary.

The confusion surrounding these issues and the ensuing criticisms convinced many that 'Heidegger's allegedly deep analysis of death does not contain significant philosophical insights. It is a mesmerising play with words, a masterly piece of rhetoric' (Philipse 1998, p. 354). Others have argued that Heidegger's death analysis is a series of 'platitudes' that are 'clearly false' and contain 'flagrant contradiction' (Edwards, 1979, pp. 50–60). I claim that these critiques stem from a failure to note two key aspects of Heidegger's death analysis. The first is the distinction between death and demise. The second is the specific sense Heidegger gives the term 'possibility' (*Möglichkeit*). I take these each in turn.

Heidegger uses three terms in his discussion of death: perishing (*verenden*), demise (*ableben*), and death (*der Tod*). Perishing indicates the end of life of any organism; all organisms perish. But Dasein perishes in a particular way, which is indicated by the term demise (Heidegger 1962, p. 291). Heidegger does not use the term 'death' to indicate the event that ends Dasein's life. This event is what Heidegger calls *demise*, which is a special case of perishing.

The ending of that which lives we have called perishing. Dasein, too, 'has' its death, the kind appropriate to anything that lives; and it has it, not in ontical isolation, but as codetermined by its primordial kind of Being [. . .] We designate this intermediate phenomenon as its 'demise' (Heidegger 1962, p. 291).

Dasein demises rather than perishes because of Dasein's way of being, *existence*, which Heidegger takes as distinct from animal life as it is endowed with self-understanding. Because Dasein *exists* (rather than merely *is*) and is engaged in an interpretation of its existence, it is aware of its future demise. But demise is not death nor dying (*sterben*), which Heidegger defines as a way of being-towards-death: '[l]et the term "*dying*" stand for that *way of Being* in which Dasein *is towards* its death' (p. 291).

Demise is commonly misunderstood as the *inauthentic* end of one's life, which stands in opposition to both authentically dying and to perishing. But, as Blattner argues, this oversimplifies the text. Heidegger is not equating demise with inauthenticity and death with authenticity, but rather claims that when Dasein understands its death inauthentically, it focuses on demise instead (1994, p. 55). Seeing demise as inauthentic death has led those who understand demise in this way to say that 'it is undeniable that there is a certain instability in Heidegger's talk of "demise"' (Mulhall 2005, p. 302).

I suggest that we should follow Dreyfus and Blattner, who take the phrase 'intermediate phenomenon' to mean that demise is intermediate between factuality (*Tatsächlichkeit*) and facticity (*Faktizität*). Perishing is *factual*, the fact that organic life is finite (in the same way that being two metres long is a fact about a particular organism). Demise is the *factical* interpretation of that fact, of which Dasein is uniquely capable (like 'being tall', a factical interpretation of 'being two metres long') (Dreyfus 1991, p. 309; compare Blattner 1994, p. 54).

Demise is not inauthentic death. Rather, when Dasein relates inauthentically to its death, it 'turns its attention to demise instead' (Blattner 1994, p. 55). When confronted with death, Dasein transforms its anxiety into fear of a future event, its demise. The event that ends life, demise, is taken as a substitute for death, because it can be dealt with by tranquillization, whereas genuine death anxiety cannot be similarly assuaged. On this interpretation death is levelled down to an ontic event because it is not understood as an *existentiale*, a way to be (Leman-Stefanovic 1987, p. 62).

By saying that Dasein is towards its death (*Sein-zum-Tode*), Heidegger picks out another feature of Dasein's existence, which gives it its finite structure. But this feature is not demise. I shall explain shortly what this feature is. First, let us look at the special sense Heidegger gives the term 'possibility'. 'Possibility' does not indicate 'a free-floating ability to be' in which Dasein is indifferent to all possibilities (Heidegger 1962, p. 183, translation modified). Nor is it a modal category, signifying what is not yet actual and not at any time necessary. Rather, possibility is an *existentiale*, a fundamental axis of Dasein's structure. Dasein is primarily 'Being possible' (1962, p. 183).

This notion of possibility emphasizes Dasein's openness and self-determination, its ability to project itself towards its future by pressing into possibilities. But as we have seen, these possibilities are limited by Dasein's thrownness and affectivity (*Befindlichkeit*). 'In every case Dasein, as essentially having an affectivity, has already got itself into definite possibilities [. . .] Dasein is Being-possible which has been delivered over to itself—*thrown possibility* through and through' (p. 183, translation modified). Therefore, Heidegger's definition of death as a possibility of impossibility requires explanation. How can death be a possibility? If 'possibility' does not mean merely logical possibility, but an *existentiale*, a way for Dasein to exist, what kind of possibility is death? Here are two answers proposed by William Blattner and Hubert Dreyfus.

7.2.1 Blattner on death

Blattner (1994) starts from the problem presented earlier, namely, how can we make sense of Heidegger's claim that death is a possibility? As we have seen, this is not just logical possibility but a possible way to exist (Heidegger 1962, §31). So saying that death is a possibility implies that there is a way of being (which Heidegger calls death) in which Dasein is unable to be. This, says Blattner, is a contradiction (compare Edwards 1975). To solve the problem one could either argue that the term 'death' does not have its ordinary meaning in *Being and Time*, or that the term 'possibility' is used in a special way in this particular formulation.

Instead of taking death to mean the end of one's life, and then changing the meaning of the term 'possibility' as others have done (Edwards 1975), Blattner redefines the concept of death. He takes the difference between death and demise to be radical and fundamental. For Blattner demise is the event that ends Dasein's life, what we usually call death, while death (in *Being and Time*) is an extreme form of an anxiety attack.

Blattner (1994) supports this view with a close reading of §52 of *Being and Time.*

We can find further support for the strict distinction between death and demise in other relevant passages of *Being and Time.* Heidegger writes: 'when Dasein dies—and even when it dies authentically—it does not have to do so with an Experience of its factical demising, or in such an experience' (1962, p. 291). Heidegger does not claim that demise (the event that ends Dasein's life) is significant for the phenomenological investigation of death, as he states explicitly and as commentators point out (Leman-Stefanovic 1987; Dastur 1996; Mulhall 2005). Additionally, as was noted earlier, the event that ends one's life, demise, is phenomenologically opaque. We cannot experience our demise or that of others (Heidegger 1962, pp. 282 and 291). *Death,* on the other hand, is, as Heidegger says, 'a phenomenon of life' and is available for phenomenological investigation (1962, p. 290; compare Heidegger 2000, p. 139).

This supports Blattner's reading of death as radically different from demise. Blattner suggests that death in *Being and Time* means an anxiety attack, the condition of being unable to be anything because one is paralyzed by anxiety. This view affords a way out of the otherwise paradoxical view that death is a possibility (or a way to be) in which Dasein is unable to be (Blattner 1994, p. 49). As Blattner says, 'death is a condition in which Dasein's being is at issue, but in which Dasein is anxiously unable to understand itself by projecting itself into some possible way to be' (1994, pp. 49–50).

Blattner takes Dasein's existence to have a thin and a thick sense. The thin sense is one of mere existence, of being an entity that has Dasein's kind of being. The thick sense is tied to Heidegger's notion of possibility as something Dasein projects itself into. The thick sense of existence is therefore that of being able to be someone by throwing oneself into some definite possibility, or understanding itself as something (1994, p. 59). In order to throw oneself into a possibility Dasein must have an affective disposition for this possibility over others, and it must stand in relation to Dasein's past choices and actions and present preferences and projects. Any particular possibility must make sense against a background of motivational preferences and understanding.

When Dasein is anxious this background disappears. Dasein's world recedes as a whole, withdrawing the framework of meaning and intelligibility that gives Dasein's preferences and projects their place and sense.

Anxiety reveals the insignificance of entities within the world, robs Dasein of its ability to understand itself and its world, and makes everyday familiarity collapse (Heidegger 1962, pp. 231–3). As a result, all possibilities become equally irrelevant for Dasein, because they lose their affective and motivational significance. As Blattner says: '*I* cease to make sense, for I am cut off from the context that lets things make sense' (1994, p. 62). Because Dasein is anxious, all possibilities become uninteresting and distant. Although Dasein still *is* in the thin sense, it is *unable to be* in the thick sense. So in anxiety Dasein is, but is unable to be (unable to press into possibilities), and this formulation matches Heidegger's characterization of death as the possibility of impossibility (1994, p. 62). Blattner thinks of death as existential death, or a moment of anxiety that removes Dasein from its world and context and bares it to itself. This makes death a limit-situation of an ability to exist in the thick sense (1994, p. 67).

This interpretation serves as a reply to critics who take the formulation of death as possibility to be incoherent. For example, Paul Edwards (1975) insists on possibility being understood as 'not actual', and therefore argues that the possibility of impossibility is simply impossibility per se.

The total absence of experiences and behaviour is most emphatically *not* what we mean by 'possibility' in any of its ordinary senses and it is *equally* not what Heidegger himself meant when he introduced the word 'possibility' in his special sense to mean the actions or conduct or mode of life which a person may choose (Edwards 1975, p. 558).

Edwards sees Heidegger's juxtaposition of possibility with actuality as 'an incomplete disjunction', because death (and prenatal non-existence) is total absence, and therefore neither possibility nor actuality (1975, pp. 560–1). But as Blattner shows, death *is* a possibility; it is existence without the ability to press into any possibility. This explains Heidegger's definition of death as a possibility. Blattner's interpretation is preferable to Edwards's because it takes Heidegger's use of the term 'possibility' to be consistent throughout *Being and Time* and moreover presents a coherent solution to the problematic formulation of death as possibility. For Blattner, the term 'possibility' has a fixed meaning, and 'death' does not mean being dead but *being unable to be*.

7.2.2 Dreyfus on death

Dreyfus, too, does not take death to mean 'the event at the end of one's life', but rather takes the term to signify how Dasein is limited. Understood existentially, death illustrates 'in a perspicuous but misleading way' that Dasein is powerless, that it can never make any possibilities its own (Dreyfus 1991, p. 310). This is because Dreyfus takes possibilities to arise from the public or social realm, which he calls 'The Background'. Because possibilities are part of the public world, they are available for anyone and therefore have no intrinsic meaning for a particular Dasein. 'They would be there whether I existed or not, and so they have no intrinsic meaning for *me*' (p. 310).

All the possibilities available to Dasein have no ultimate reason or grounding in Dasein. Dasein's preferences are to an extent arbitrary, because they arise from Dasein's social environment and are defined by its thrownness. Therefore, the only possibility that essentially belongs to Dasein is its nullity or groundlessness. So death, claims Dreyfus, is not the *existentiell* or ontic possibility of demise, but the *existentiale* ontological possibility of not having any possibilities. This extreme existential nullity is covered over by thinking about death as an event that has not yet happened to me, rather than as the condition of my existence. 'The cover up consists in assuming that the anxiety of death is a response to the end of being alive or to the possibility of that end rather than to the true condition of Dasein' (1991, p. 311).

Both Blattner's and Dreyfus's views are supported by Heidegger's identification of being-towards-death and anxiety: 'Being-towards-death is essentially anxiety' (Heidegger 1962, p. 310). Anxiety is the affective state (*Befindlichkeit*) in which Dasein discloses its groundlessness, its inability to project itself into a possibility, or its inability to be. Dreyfus thinks that anxiety reveals the publicness of possibilities: that they belong to anyone and therefore cannot provide a source for Dasein's fixed identity (1991, p. 310). Blattner similarly says that in anxiety Dasein is cut off from its possibilities, and is therefore unable to be in a thick sense.

These possibilities, which are usually taken up unreflectively, provide a transparent, shared backdrop to everyday life. Once these possibilities are removed, like in anxiety, the backdrop of intelligibility disappears as well, leaving Dasein unable to project itself into any particular possibility, or in Blattner's words, unable to *be*. Dreyfus and Blattner emphasize the inability to act that characterizes anxiety and what they call 'existential death' (Dreyfus 1991, p. 310; Blattner 1994, p. 68).

It is this condition of being cut off from the world and therefore being incapable of action that death and anxiety share. This unique condition is what Heidegger calls non-genuine authenticity. The terms genuine (*echt*) and non-genuine (*unecht*) pertain to understanding (Heidegger 1962, p. 186). Genuine understanding expresses being-in-the-world as a whole, whereas non-genuine understanding is partial or reductive. Anxiety or authentic being-towards-death, qualifies as *authentic* but *non-genuine*. These states are authentic because they disclose the world as a whole, but they are non-genuine because they cut Dasein off from its world and leave it unable to act (Dreyfus 1991, p. 194).

The condition of anxiety, or of being unable to be, is a condition of non-genuine authenticity. When it is anxious, Dasein is equipped with authentic understanding but unable to enact it. In order to achieve authentic *and* genuine understanding Dasein must be resolute, which allows it to act with a 'sight which is related primarily and on the whole to existence', which Heidegger calls transparency or perspicuity (*Durchsichtigkeit*) (1962, p. 186). Death and anxiety are both conditions in which action is ruled out, in which Dasein is unable to be.

This is a compelling account of Heidegger's concept of death that solves the apparent contradiction in defining death as possibility, and accounts for the difference between death and demise. Dreyfus and Blattner understand Heidegger's concept of death as the breakdown of Dasein's world, which makes Dasein unable to be anything (Dreyfus 2005, pp. xix–xx). More recently Dreyfus has expressed the view that death is the structural condition of a complete loss of identity. He interprets dying as the 'resigned, heroic acceptance of this condition' (pp. xix–xx). This formulation is close to Blattner's idea of death as being unable to be anything; or in Dreyfus's terms, losing one's identity.

Dreyfus's and Blattner's view offers the most coherent interpretation of the difficult passages on death in chapter I of Division II of *Being and Time*. What I refer to from now on as the Dreyfus/Blattner view is the following: death is the condition of being unable to be anything, when being is taken in the thick sense of pressing into a possibility. This continuous threat of loss of identity and anxiety structures Dasein's existence. When Heidegger discusses death, being-towards-death, and anxiety, he is not referring to demise or to our attitude towards demise, but to this condition.

7.2.3 Problems with the Dreyfus/Blattner view

Despite its ability to provide a coherent interpretation of the difficult passages of *Being and Time*, the Dreyfus/Blattner view faces some difficulties. A first problem is pointed out by Dreyfus himself, who criticizes Blattner. He argues that since an anxiety attack is sudden and unmotivated 'it is hard to see how one should live in order to be ready for it, and Blattner does not even try to explain what a life of readiness for an anxiety attack would be like' (2005, p. xix).

Additionally, it is not clear that Heidegger thinks that Dasein can be ready for this kind of attack, and moreover, anticipatory or forerunning (*vorlaufen*) resoluteness is already constantly anxious (Drefus 2005, p. xix). If authentic Dasein is constantly anxious, it does not need to be ready for an anxiety attack. Dreyfus suggests that Blattner's notion of an anxiety attack is in fact not an authentic experience of death, but the nearest experience to death that an *inauthentic* Dasein can have (p. xix).

There are further problems with the Dreyfus/Blattner view. Their interpretation solves local problems in specific sections of *Being and Time*, at the expense of the overall meaningfulness of the concept of death. Their position lacks the notion of *temporal* finitude, without which many central ideas of *Being and Time* cease to make sense. The analysis of temporality and the phenomenology of our attitudes towards the death of others, the characterization of death as certain, and the analysis of other types of ending, are all central elements of *Being and Time* that require the notion of temporal finitude.

Thus in the Dreyfus/Blattner interpretation central themes of *Being and Time* lose their connection with the concept of death, which has lost its temporal dimension. The concepts surrounding temporality lose their coherence and relevance to the main thrust of Division II if we understand death through the Dreyfus/Blattner interpretation. The Dreyfus/Blattner view solves the problems that obfuscate Heidegger's discussion of death, but at the same time their analysis is incomplete because it lacks the notion of temporal finitude that is so crucial to Division II. I now suggest an augmented interpretation.

7.2.4 Amending the Dreyfus/Blattner interpretation

In order to provide the missing dimension of temporal finitude to Heidegger's concepts of death I augment the Dreyfus/Blattner view. This provides the link between death as finitude of possibility (the Dreyfus/Blattner interpretation) and temporal finitude. On my understanding, death illuminates both types of finitude: finitude in possibility and temporal finitude. This essential element makes coherent the emphasis on temporality, historicality, and finitude and links these terms to the analysis of death in a way that makes *Being and Time* a coherent whole. I therefore suggest we understand the term 'death' in *Being and Time* as containing both temporal finitude and finitude of possibility.[10] The two types of finitude are internally related, and both enable the transition to authenticity.

There are several reasons why this addition is necessary. First, Heidegger's discussion of being-towards-the-end sees Dasein's mortality as a structuring principle and central concern for Dasein (compare Leman-Stefanovic 1987, p. 6). Being-towards-the-end defines Dasein as finite temporality, as a constant projection towards its future annihilation. Temporal finitude underlies all possible projections into the future. As such, Dasein's death as its temporal end is a constant condition of every possibility. Death is constantly present in life because it is only ever impending and can never be actualized, that is, experienced by Dasein.

Moreover, whereas other things are possible only at certain times, Dasein's end is possible at every moment. Our end is 'always and only a possibility' and as such is persistently present in Dasein's existence (Mulhall 2005, p. 303). Second, temporal finitude provides further illumination of the formulation of death as a possibility. In the same way that on the Dreyfus/Blattner view death was not a possibility in the ordinary sense of the word, but a possibility of being unable to take up any possibility, Dasein's temporal end is also not a possibility in the ordinary sense.

Dasein's end is not a possibility waiting to be realized, but an ontological condition of Dasein's temporal structure. Both the possibility of being unable to be and temporal finitude are conditions for the

---

[10] Richard Polt suggests the term 'mortality' (1999, p. 86). The problem with this term is that it does not designate the end of Dasein's life, but the condition of being finite.

meaningfulness of all other possibilities and both are limit cases that define the boundaries of meaningful experience. Dasein's finitude of possibility is an ever-present threat to Dasein's everyday meaningful dealing with the world. Dasein's finitude is a fundamental aspect of its existence that accompanies every moment of life. The result is a being that is both mortal and finite.

Third, finitude of possibility and temporal finitude are internally related. Both define the limits of life, and as limit concepts they assign significance to life by delineating it. Being-towards-the-end requires temporal finitude; death (in the Dreyfus/Blattner sense) is the finitude of possibilities. The two are also related through the concept of anxiety. Anxiety is the state of being dead in the Dreyfus/Blattner sense; it is also the affective state that discloses Dasein to itself as being-towards-the-end. Both types of finitude raise issues that do not pertain to demise, but to the question of how one ought to live knowing that one will die. In other words, the concept of death reformulates human life as mortal and finite.

Dasein's relation to death is not something that is realized when it demises, but something it either realizes or fails to realize in its life (Mulhall 2005, p. 303). Confronting death illuminates that Dasein's being is always an issue; that 'its life is something for which it is responsible, that it is its own to live (or to disown)' (2005, p. 306). Because death is possible at any moment, the radical contingency of life becomes apparent. Acknowledging this is to acknowledge finitude in three interrelated senses: the groundlessness of Dasein (Dreyfus), the possibility of impossibility (Blattner), and temporal finitude. As Mulhall says, death makes explicit the fact that our existence has limits, 'that it is neither self-originating nor self-grounding not self-sufficient, that it is contingent from top to bottom' (Mulhall 2005, p. 306; compare Hatab 1995, p. 411).

We now have a complete picture of death as both temporal finitude *and* finitude of possibility. Death is a limit shaping life; we are finite and mortal. The shape we give our life, therefore, depends on our attitude to finitude, which shapes Dasein's temporality (*Zeitlichkeit*). Temporality is the meaning of Dasein's totality, or care (Heidegger 1962, pp. 425–6). Dasein is a temporal and historical being, embedded in a community. The temporal meaning (*Sinn*) of Dasein is *Mitsein*, a social and historical being-with. It is impossible to regard Dasein as a sum total of past and present events, because this reduces it to an object with fixed properties, existing within time as a series of 'nows'. Against this view, Heidegger defines Dasein as temporal change and movement (*Bewegtheit*), which articulates the gap between Dasein and its possibilities and the stretch between Dasein's birth and death:

> Factical Dasein exists as born; and, as born, it is already dying, in the sense of Being-towards-death. As long as Dasein factically exists, both the 'ends' and their 'between' *are*, and they *are* in the only way which is possible on the basis of Dasein's Being as *care*. Thrownness and that Being towards death in which one either flees it or anticipates it, form a unity; and in this unity birth and death are 'connected' in a manner characteristic of Dasein. As care, Dasein is the 'between' (Heidegger 1962, pp. 426–7).

This gap never closes because the projection into possibilities never halts and because Dasein is never pure possibility; it is always also thrown. Dasein is never united with the 'there', the world or its possibilities, and is therefore always projecting towards it. Dasein understands itself not only as 'I am' but also as 'I can be' and 'I already am', as having a dynamic openness. The disclosure of possibilities is enabled through temporal existence drawing us from the past to the future, allowing us to see ourselves as an ability-to-be immersed in a given world, as thrown projection.

In primordial temporality the projective character of Dasein is 'ahead of itself', grounded in the future, which expresses Dasein's *existentiality* (Heidegger 1962, p. 376). But Dasein always also finds itself as a thrown fact, as *facticity* (p. 376.). The third temporal mode, the present, is included in the past and the future, either as *falling*, which is the inauthentic present, or as resolute. When resolute, Dasein *brings itself back* from falling in order to be authentically 'there' in the moment.[11] The unity of existence (characterizing the future), facticity (characterizing the past), and falling (or bringing oneself back from falling as the present) constitutes the totality of the structure of care.

---

[11] 'The moment' (Dreyfus) is a translation of *Augenblick*, also translated as 'the instant' (Hofstadter), or 'the moment of vision' (Macquarrie and Robinson). I shall follow Dreyfus because this formulation lends itself nicely to such expressions as 'being in the moment' and 'the moment of transformation', as well as being consistent with the English translation of Kierkegaard's *Oieblik*, which is the source of Heidegger's term (Dreyfus, 1991, p. x).

Finally, originary temporality is *finite*. The unity of past and future is contained within the present. Falling (*verfallen*), the present absorption in the world and everydayness, is not a negative state (Heidegger 1962, p. 220). It represents a normal state of affairs in which Dasein plans and performs tasks while immersed in idle talk, ambiguity, and curiosity. Falling is Dasein's basic kind of existence that belongs to everydayness (p. 219). It is not a 'bad and deplorable' ontic state Dasein should aim to get rid of (pp. 220–1). The past enables the future to disclose itself as a horizon of new possibilities. In authenticity the present is articulated through resoluteness, which is an attitude towards new possibilities. The present possibilities are the ontological source of the present, which is the relation between past and future (p. 351).

Dasein always projects itself towards its death, and therefore every temporal dimension of Dasein contains mortality as its boundary. Temporality is the axis through which finitude becomes meaningful, since the movement towards the future is also the movement towards the end. It is only at the meeting point of temporality and historicality, when temporality is instantiated in entities as historicality, that Dasein can become authentic. Repetition enables authentic historicality, by explicitly handing down Dasein's possibilities, connecting past and future. This historicality is based on Dasein's understanding of itself as temporally finite, as being-towards-death (Heidegger 1962, p. 438).

Inauthentic temporality, on the other hand, is the projection of a substantial structure onto this dynamism. In inauthentic temporality we find repetition replaced by *forgetting*, the moment by *making present*, and anticipation by *awaiting* (Heidegger 1962, §68). Authentic temporality is not merely a structural horizon of Dasein but an active process of selection of factical contents, determining those that are to be preserved and repeated. Temporality is of course related to mortality. Temporality always structures Dasein as mortal, even if Dasein fails to understand itself as finite. Finitude puts certain restrictions on Dasein's existence but also gives Dasein its temporal unity, as well as the context within which it can receive its heritage and ground itself as historical.

The individuation experienced in authentic being-towards-death is also present in illness, because illness punctures intersubjectivity. The shared world, which is based on shared embodied experiences and practices (e.g. Dreyfus's Background) is fundamentally disrupted in illness. On Heidegger's reading, this disruption is necessary in order to

loosen *das Man*'s grip on Dasein and make room for authentic resoluteness. But is individuation an essential feature of death, as Heidegger
claims, and of illness? Or could a different understanding of the
relationship between authenticity and inauthenticity lead to a more
solicitous experience of illness and dying? The final section of this
chapter suggests just this.

7.3   Authentic Attitude to Death and Illness

Authenticity and inauthenticity are basic possibilities of Dasein's existence
(Heidegger 1962, p. 235). Heidegger also introduces a third mode
of indifference (*Indifferenz*). The three modes stem from 'mineness'
(*Jemeinigkeit*), the ontological condition of these ontic modalities (p. 78).
Authenticity and inauthenticity occupy opposite ends of a polarity, yet
both belong to Dasein (Nancy 1993, p. 100; Heidegger 1962, p. 68). Both
modes are conditioned by a claim to a first-person existence, whose
relation to the public sphere could be harmonious, oppositional, or
one of subjection. Both are positive and concrete ways of taking up
possibilities, and as such are grounded in being-in-the-world.

We live in the mode of indifference until we hear the call of
conscience. This can lead Dasein to authenticity or inauthenticity,
depending on its decision as to whether to answer the call or ignore it.
In the mode of indifference the call has either gone unnoticed or never
took place. The call is expressed through reticence, and is contrasted with
the chatter of the 'they-self' (*das Man-selbst*). 'The they-self keeps on
saying "I" most loudly and most frequently because at bottom it is not
authentically itself [...]. As something that keeps silent, authentic
Being-one's-Self is just the sort of thing that does not keep saying "I"'
(Heidegger 1962, pp. 369–70).

In the inauthentic mode Dasein is taken over by *das Man*, covering up
its death, in both the Dreyfus/Blattner sense and as temporal finitude.
Existentialist readings of *Being and Time* endorse the notion of an
authentic self that has been corrupted by *das Man*, and therefore regard
authenticity as a liberating achievement. But pragmatic interpretations
see *das Man* as the agency of social norms, which provides the basis for
intelligibility. Dasein needs *das Man* as the condition for communication
and sharing a world with others. On this reading, pure authenticity is
impossible because of the essential role of *das Man*.

The difficulty in determining the relationship between the two modes comes from the text, in which Heidegger makes two contradictory statements. At the end of §27 of *Being and Time* Heidegger claims that authenticity is a modification of inauthenticity: 'Authentic Being-one's-self does not rest upon an exceptional condition of the subject, a condition that has been detached from the "they"; it is rather an existentiell modification of the "they"—of the "they" as an essential existentiale' (Heidegger 1962, p. 168). But in §64 he claims the opposite: 'It has been shown that proximally and for the most part Dasein is not itself but is lost in the they-self, which is an existentiell modification of the authentic self' (Heidegger 1962, p. 365).

To explain this seeming contradiction, some argue that authenticity is ontological and therefore primary, and that inauthenticity is a fall from authentic grace (Stambaugh 1978). Others claim that inauthenticity must be an original mode, and therefore inauthenticity itself cannot be a modification of authenticity (Macann 1992, p. 233). Carman argues that both are ontic states (1994, p. 217). Lewis claims that authenticity is impossible (2005, p. 35). As a result of these interpretative difficulties some commentators think that the distinction does not hold at all.

Another line of debate regards the possibility of an authentic everyday. Is authenticity a momentary state or a continuous mode of existence? Some authors, such as Rudi Visker (1996), regard authenticity as a momentary leap out of inauthenticity, and claim that as such it must recur. Visker criticizes Heidegger for promising to provide an authentic everydayness, a promise that cannot be met. Because he views authenticity as momentary, Visker thinks that in order to have basic continuous existence, we must be inauthentic (Visker 1996, p. 80).

Visker overlooks two important points. First, authenticity need not be thought of as momentary. Anxiety is momentary but the authenticity that follows is a return to the everyday, to the full thrust of thrown projection and worldly life. Recall the distinction between genuine and non-genuine authenticity. Anxiety is non-genuine authenticity because in it Dasein does not fully participate in its world. Non-genuine authenticity is a momentary anxiety attack, but the genuine authenticity that may follow is a continuous mode of existence.

Second, authenticity does not arise solely from inauthenticity but may also arise out of indifference, in which case its relationship to inauthenticity remains unclear. It is easy to conflate inauthenticity with indifference but the two are clearly distinguished by Heidegger (1962, p. 69, p. 78). The undifferentiated mode is average everydayness, prior to the call of conscience. After the call, the response could lead to either authenticity or inauthenticity but Dasein could also remain undifferentiated, if the call is not heard.

Another reason that should lead us to give a central place to inauthenticity and indifference is that the transition to authenticity is individual. While one Dasein becomes authentic, everydayness and *das Man* remain. But does that mean that the everyday is completely subsumed under *das Man*? According to Visker, the answer is yes. If we remain in the everyday we remain forgetful, hence there is no possibility of genuine authenticity which is immersed in the everyday world.

Each of the above interpretations confronts difficulties of internal consistency as well as being textually selective. I propose to solve the interpretative impasse by rejecting the claim that the distinction between authenticity and inauthenticity is clear-cut. On this reading, the contradictory statements Heidegger makes require us to reformulate the distinction as blurred and the two states as internally related. I suggest that this view enables us to view authenticity as immersed in everydayness, and as a state that can only initially arise out of either the undifferentiated mode or inauthenticity. The textual evidence cited above supports this view of the modes as interdependent. If we no longer understand authenticity and inauthenticity as dichotomous, then the move to authenticity does not entail leaving *das Man*, everydayness, or being-with-others and does not lead to the ensuing interpretative difficulties. Rather, authenticity becomes the full disclosure of the horizon of the everyday, fully retaining Dasein's relationship to other Dasein and to the world. This includes the everyday structures of intelligibility that seem to depend on *das Man*.

I suggest that, rather than resigning the everyday to inauthenticity, we should dismantle the ideal of pure authenticity by regarding inauthenticity and indifference as necessary components of existence. This position allows the everyday to have a significant disclosive function, because it allows Dasein to be authentic as a way of being-in-the-world. Intelligibility is given in the world, in everyday existence. Falling is not a descent from authentic existence to an inferior state, but a ground in itself (Heidegger 1962, p. 68; compare Nancy 1993, p. 99). Dasein's possibilities are disclosed through everydayness and averageness and do not exist beyond it. Therefore, inauthentic everydayness is the source of intelligibility.

This position sees authenticity and inauthenticity as equally primordial; both are ontic modalities of indifference. This enables a new understanding of the two states: the two modes of existence are seen as mutually dependent. This reading resolves the contradictory nature of the authenticity/inauthenticity distinction, thereby opening everydayness to its disclosive capacity.

This view solves a further problem, raised by Dreyfus (1991). Dreyfus distinguishes two accounts of falling. One account sees falling as absorption in equipment and conforming to *das Man*. On this view levelling goes along with practical intelligibility. The second account is of Dasein actively resisting the call of conscience. Here levelling is not structural (falling) but a motivated form of covering up (fleeing). Dreyfus argues that this makes inauthenticity both inevitable and incomprehensible (1991, p. 333ff.). It is inevitable because absorption in the present is inevitable. It is incomprehensible because if authenticity is so rewarding why would Dasein return to inauthenticity?

Dreyfus claims that Heidegger does not address these questions and takes the fact that Heidegger ceases to talk of fleeing thereafter as indicating that he recognized the problem. If we view the everyday as undifferentiated rather than inauthentic, and introduce the possibility of authentic everydayness, inauthenticity would cease to be inevitable. Thus the integrated view responds to the first of Dreyfus' criticisms. As for the second criticism, that inauthenticity is incomprehensible once authenticity is achieved, we find a compelling reply in Carman. He argues that resoluteness is not a stable, self-sufficient mode of existence, but a perpetual struggle against the levelling and banalizing forces of idle talk, ambiguity, and curiosity. 'Authentic existence is thus constituted by the very forces against which it has to push in its effort to grasp itself in its facticity' (Carman 2000, p. 24). Without the forces of levelling and tranquillizing, authenticity would be meaningless; it is only meaningful as a position of resistance and refusal. Therefore the threat of inauthenticity is perpetual.

Dreyfus's further question—why it is difficult to maintain authentic resoluteness—is answered by looking carefully at the language with which authenticity and resoluteness are described. The choice of authenticity is not a choice at all, but a *gestalt* switch or transformation that comes from Dasein accepting its powerlessness (Dreyfus 1991). It is a new form of self-understanding and formal view of the world, not a specific reply to moral deliberation. Authenticity is not expressed in the specific contents of Dasein's choices, but in the expansion of its ability to view itself and its world and to respond to this new perspicuity. Authenticity is a new openness or ability to view oneself as a whole; it is a structural shift.

This interpretation is further supported by looking at the German term *Entschlossenheit* (resoluteness). As both Dreyfus and Albert Hofstadter point out, *Entschlossenheit* means determination or resolve, but *Ent-schlossenheit* means unclosedness, openness. It is not so much an expression of Dasein's will or choice, as openness or letting be. Heidegger says in *Introduction to Metaphysics*:

> To will is to be resolute. The essence of willing is traced back here to open resoluteness. But the essence of open resoluteness <*Ent-schlossenheit*> lies in the de-concealment <*Ent-borgenheit*> of human Dasein for the clearing of Being and by no means in an accumulation of energy for 'activity'. But the relation to Being is letting (2000, pp. 22–3).

In 'The Origin of the Work of Art' Heidegger writes, 'The resoluteness intended in *Being and Time* is not the deliberate action of a subject, but the opening up of human being, out of its captivity in that which is, to the openness of Being' (1993, p. 67). Inauthenticity is the constant temptation of fleeing that is part of everydayness and is present as a threat even in authenticity. Falling is a starting point and not a fall from authenticity to inauthenticity. As a result, inauthenticity assumes a grounding function as a continuous possibility against which authenticity is defined.[12] Lawrence Vogel (1994) makes the same point, saying that possibilities do not come from nowhere but are handed down to Dasein from the factical world to which it belongs. '*Das Man* does not arise from external circumstances but from the heart of Dasein's existence' (1994, p. 12).

A further misunderstanding lies in the assumption that authenticity requires isolation, or that it clashes with *Mitsein*. This identifies

---

[12] Abraham Mansbach supports this interpretation: 'total authentic existence cannot be achieved, for the "they" is an original constituent element of factical Dasein [...] [Inauthenticity] belongs to Dasein's essential nature, it is a mode of being-a-self' (1991, p. 81). See also Einar Øverenget: '[Authenticity] does not appear prior to and independently of our dealings with the world' (1998, p. 259).

inauthenticity with *Mitsein* and authenticity with solitude. But authenticity does not entail solitude. It is true that when faced authentically, death individuates Dasein and anxiety cuts it off from its world. But this is only the first step, that of non-genuine authenticity. Genuine authenticity is inherently linked to *Mitsein*. As Heidegger writes:

Resoluteness, *as authentic Being-one's-Self*, does not detach Dasein from its world, nor does it isolate it so that it becomes a free-floating "I". And how should it, when resoluteness as authentic disclosedness, is *authentically* nothing else than *Being-in-the-world*? Resoluteness brings the Self right into its current concernful Being-alongside what is ready-to-hand, and pushes it into solicitous Being with Others (1962, p. 344).

As opposed to anxiety, resolute existence contains both being-alongside and being-with. Therefore resoluteness contains a high level of communal commitment and involvement with others. Carman argues in support that authenticity does not mean consisting entirely of first-person perspective. Not all overlapping of first- and third-person perspectives amounts to a loss of self and alienation (2005, p. 287). The ability to perceive oneself from second- and third-person perspectives is an important part of being-with. So authenticity cannot merely mean being wholly oneself as simply removing all other points of view from a person's self-understanding.

How does this integrative interpretation affect our understanding of death (in both the Dreyfus/Blattner sense and as temporal finitude) and of illness? Authenticity is a relationship of struggle and conflict with death and finitude, while inauthenticity covers over mortality and the possibility of anxiety. The same struggle and conflict may characterize an authentic relationship to illness. Inauthenticity offers psychological comfort at an epistemic price: it covers over the truth of our groundlessness and mortality in a life of forgetful existence. Resolutely facing death is the condition of authenticity.

Once the tension between the two is dissolved, as I propose here, the emphasis on individuation is no longer required and a path is opened to a more compassionate experience of illness and finitude. The emphasis on authenticity as individuation, as retreating from the world and severing the relations to it, is removed. Authenticity becomes a genuine engagement with the everyday world, not its erasure. Therefore individuation is no longer a crucial factor for death and authenticity, or a compulsory element of illness. Although Blattner's anxiety attack is individuated and therefore non-genuine authenticity, it is followed by genuine authenticity, an authentic form of being-with and full participation in the world.

Death as temporal finitude also does not require a complete breakdown of Dasein's relations. Against Heidegger's emphasis on *my own* death as the only way to understand mortality, I suggest that grief and mourning for others can also intimate mortality and thus have an edifying role. The emphasis on resoluteness as solicitude and on the community of authentic Dasein becomes more tenable on this relational reading of authenticity. Everydayness can underpin a sustained social existence, which enables us to link authenticity and *Mitsein*. This also supports the idea that the death of another is a disclosive experience. Finally, anxiety loses its exclusive role as the only affective state leading to authenticity.

In the case of illness, the two kinds of finitude are present as well. Illness limits possibilities more concretely than death. It also intimates mortality and temporal finitude. But the Heideggerian requirement for individuation seems even more redundant in the case of illness. In order to face illness authentically, Dasein does not need to sever its links to the world. An authentic attitude to illness may include resolutely facing illness, refusing to repress its impact, and accepting its presence.

This authentic attitude may be difficult to cultivate in public contexts—i.e. support groups, or online patient fora—but it is still made possible by a close community of friends or other patients who are attuned to the possibility of accepting illness and existing with it without denial or fleeing. Our understanding of illness stands only to gain by this integrative reading of authenticity and inauthenticity offered in this chapter. As I have shown, there are good reasons to reject the sharp divide between authenticity and inauthenticity and to claim that an authentic attitude to the death of the other is possible. Similarly, illness can be experienced through solicitude and illness can be confronted in a way that is both authentic and shared.

7.4 Conclusion

This chapter offered an account of death and of its relation to illness through a detailed reading of Heidegger. It provided an interpretation of the difficult passages in *Being and Time* that describe Dasein as being towards death, and discussed several attempts to resolve the textual difficulties that caused much confusion in the secondary literature. It suggested that the Dreyfus/Blattner view, that death is the condition of being unable to be anything, when being is taken in the thick sense of pressing into a possibility, needed augmenting. It then offered to amend their view by adding a second meaning, that of temporal finitude, to their notion of death as finitude of possibility. Finally, the chapter criticized the reading of Heidegger that claims that the authentic facing of death demands individuation, and suggested that a more relational under- standing of death, and of illness, would be more productive and more faithful to our own experiences.

8

Epistemic Injustice in Healthcare

So far this book explored the first-person phenomenology of illness, and how it challenges common assumptions about illness. Now I would like to turn to the question of how such first-person reports are received by others. Are they met with belief and trust, or with suspicion and disbelief? Evidence from social psychology, health economics, patient organizations, and healthcare research shows that what patients say is more likely to be misunderstood, ignored, or rejected than reports from other people.

In this chapter we suggest that this set of reactions to illness accounts and patient reports constitutes what philosopher Miranda Fricker calls epistemic injustice; a form of injustice that is uniquely epistemic, i.e. is done to the speaker in their capacity as *knower*. Epistemic injustice is caused by biases and negative stereotypes about illness that can lead interlocutors to treat ill persons' reports with unwarranted disbelief or dismissiveness. We suggest that such treatment amounts to epistemic injustice, and examine the nature of such injustice in the context of illness.

'That really, really hurts', says the woman who has just given birth.[1] The doctor is sitting by her splayed feet, which rest on stirrups either side of him. He is stitching her vagina, his face inches away from her body. A crowd of doctors and nurses surround the baby lying a few feet away. He is being bundled, rubbed, and his airways cleared by eight health professionals. None of them seem to hear, or respond to, the woman's

[1] This chapter was coauthored with Ian James Kidd, and I therefore use 'we' throughout. It is reprinted here with kind permission from Springer Science+Business Media: *Medicine, Healthcare and Philosophy*, 'Epistemic injustice in healthcare: a philosophical analysis' 17(4), 2014, pp. 529–40, Havi Carel and Ian James Kidd.

complaint. She repeats: 'That hurts. Are you using anaesthetic?' 'No', the doctor replies calmly, 'there is no need to. I'm nearly finished'. The woman is too exhausted to persist and says nothing more. It is hard to imagine another situation in which we would not offer pain relief to someone having a needle pushed through their genitals. But the woman's testimony is not acted upon. Her pain is either not fully registered or not considered worthy of response.[2]

Similar situations arise in the context of healthcare provision. Many of us are familiar with stories about doctors who don't listen, large-scale healthcare systems that are impersonal and bureaucratic, and feelings of helplessness when trying to navigate these systems (for example Beckman and Frankel, 1984; Korsch et al. 1968, 1969). Many authors (for example Kleinman 1988; Toombs 1987) have drawn attention to the epistemic aspects of these complaints, and this work informed subsequent changes to healthcare policy in the UK, such as the NHS Patient Charter and the more recent NHS constitution.

Despite this greater awareness, patients continue to voice epistemic concerns, which attest to persistent experiences of being epistemically marginalized or excluded by health professionals (see Frank 2010; Carel 2013a). The UK's Patients Association, for instance, lists 'communication' as one of the four most frequent complaints received by the association.

Focusing on the epistemic dimension of these situations, we suggest that patients' testimonies are often dismissed as irrelevant, confused, too emotional, unhelpful, or time-consuming to deal with. A common complaint from clinicians is that patients provide irrelevant information, that patients are (understandably) upset and therefore can be irrational, and that listening for medically relevant information precludes listening to other information conveyed in patient speech, such as existential concerns, need for empathy, or emotional content.

In addition, since patients are not properly trained in the relevant medical terminology and the particular discourse of health professionals, anything they did say may be judged to be insufficiently articulate.[3] So even if the patient's testimony were relevant, emotionally balanced,

---

[2] I observed this scene while shadowing a paediatrician consultant at a UK hospital in 2008 (details redacted to ensure patient confidentiality).

[3] These discourses may vary greatly; we are not suggesting that there is only one such discourse.

and so on, what they say is not expressed in the accepted language of medical discourse and may therefore be assigned a deflated epistemic status. As one physician said: 'patients say a lot of irrelevant things like "when I eat lettuce my elbow hurts". I have to listen carefully for the important stuff and ignore the rest' (personal communication).[4]

We propose to submit this problem to an epistemic analysis, using Fricker's (2007) notion of epistemic injustice. We argue that ill people are more vulnerable to testimonial injustice, because they are often regarded as cognitively unreliable, emotionally compromised, or existentially unstable in ways that render their testimonies and interpretations suspect. We present some examples involving both somatic and mental illness. Ill people are also more vulnerable to hermeneutical injustice, because the kind of experiences illness affords are often difficult to make sense of and communicate (Carel 2013a). Perhaps certain extreme and unique experiences cannot be communicated in any direct, propositional manner, and so are only shareable with persons with whom one shares a standpoint or a sense of solidarity.[5]

We further argue that health professionals are considered epistemic-ally privileged in both warranted and unwarranted ways, by virtue of their training, expertise, and third-person psychology. Moreover, they decide which patient testimonies and interpretations to act upon. We contrast cases in which patients are assigned undeservedly low credibility with cases in which patients' credibility is undeservedly high. We show that these are two ways in which health professionals' clinical judgement can be skewed as a result of mis-assigning credibility to patients. In certain extreme cases of paternalistic medicine, patients might simply not be regarded as epistemic contributors to their case in anything except the thinnest manner (e.g. confirming their name or 'where it hurts'). Denying someone credibility they deserve is one form of epistemic injustice; denying them the role of a contributing epistemic agent at all is a distinct form of epistemic exclusion (Hookway 2010).

---

[4] Perhaps it is also a sort of epistemic injustice to complain that a person's style of testimony is no good (inarticulate, etc.) but do nothing to ameliorate this (e.g. by critically reflecting upon the reasons that one has for using these unhelpful formats rather than others). Epistemic injustice might arise because (a) one buys into epistemically unjust structures or because (b) one fails to challenge those structures.

[5] For an engaging discussion of the relationship between epistemic injustice, standpoint, and solidarity, see Medina (2012).

Finally, we suggest that the structures of contemporary healthcare practice encourage epistemic injustice because they privilege certain styles of articulating testimonies, certain forms of evidence, and certain ways of presenting and sharing knowledge, e.g. privileging impersonal third-person reports, in ways that structurally disable certain testimonial and hermeneutical activities. To address this problem we propose that phenomenology, and in particular a phenomenological toolkit (Carel 2012), may provide a useful hermeneutic context within which patients can reflect on and share their illness experiences with other patients and health professionals.

The toolkit (and similar reflective practices) may improve communication because patients benefiting from the toolkit would be better able to articulate their experiences and thereby be more effective contributors to their care. Similarly, health professionals benefiting from the toolkit would have a more nuanced grasp of patients' illness experience, as well as honing their epistemic sensibilities and skills, such as listening to and understanding multiple perspectives. The phenomenological toolkit can address hermeneutical injustice by providing patients with a framework through which to understand their experiences. It can address testimonial injustice by both helping patients articulate their illness experience and aiding health professionals in understanding it.

The structure of the chapter is as follows: section 8.1 presents Fricker's notion of epistemic injustice and discusses illness as a case of testimonial and hermeneutical injustice. Section 8.2 provides examples from healthcare to support our claims. Section 8.3 outlines the epistemic privilege of health professionals and how healthcare practices are structurally disabling. Section 8.4 puts forward the phenomenological toolkit as one type of remedy for the problem of epistemic injustice in healthcare.

8.1 Testimonial and Hermeneutical Injustice

Fricker argues that 'there is a distinctively epistemic kind of injustice' which is a wrong done to someone in their capacity as knower (2007, p. 1). She identifies two such wrongs: testimonial injustice and hermeneutical injustice. Testimonial injustice occurs when prejudice causes a hearer to assign a deflated level of credibility to a speaker's testimony. Hermeneutical injustice occurs when a gap in collective interpretative resources puts a speaker at a disadvantage when trying to make sense of their social experiences (p. 1).

We suggest that these two kinds of injustice characterize many attitudes ill people encounter when they try to voice their opinions about their care, convey their experiences, or state their priorities and preferences. In particular we suggest that an ill person may be regarded as cognitively unreliable, emotionally compromised, existentially unstable, or otherwise epistemically unreliable in a way that renders their testimonies and interpretations suspect simply by virtue of their status as an ill person with little sensitivity to their actual condition and state of mind.

Epistemic injustice can manifest in different ways and to different degrees. Although all instances of epistemic injustice share a common conceptual core, that of downgrading certain persons' testimonies and interpretations, the precise forms of such injustice can vary greatly, ranging from the blunt and brutal to the subtle and difficult to spot. In the case of illness, the forms that epistemic injustice can take will be shaped by a range of factors, including particular healthcare policies, medical practices, and even the format of patient feedback forms. Such issues clearly make the task of identifying and characterizing the epistemic injustice that ill persons experience much more difficult, but they also offer a double advantage.

First, they make it easier to identify the specific practices and policies that generate epistemic injustice in a given case. Second, they help to pinpoint our claim: it is not that modern healthcare practices are epistemically unjust but that certain policies, practices, and cultural norms within modern healthcare practice are liable to generate epistemic injustice. We do not suggest that the patient–clinician relationship is necessarily and inevitably an epistemically unjust one, but rather that certain forms it can take are prone to generate epistemic injustice. We aim to offer a means of identifying practices and biases that lead to epistemic injustice in healthcare practice rather than to criticize this practice *tout court*.

The charge of epistemic injustice can be analysed in terms of the two specific forms which Fricker identifies: *testimonial injustice* and *hermeneutical injustice*. Although some writers have identified other forms that epistemic injustice can take, we take it that these are supplements to, rather than rejections of, the twin concepts of testimonial and hermeneutical injustice as Fricker defines them.[6]

---

[6] See for instance Coady (2010) and Hookway (2010).

In addition, there are cases in which the clinician's epistemic authority is warranted but where epistemic damage is incurred nonetheless. These include cases where the style of interaction between clinician and patient is one that closes down communication, such that important information is potentially lost. In this chapter we focus on cases in which the clinician's assumption of epistemic authority over a patient in relation to matter x is mistaken. But we might also find that even when the clinician's assumption of epistemic authority in relation to matter x is correct, the clinician's style of interaction is overly dismissive.[7] Her disregard of the patient's perspective on x might still be detrimental to the patient's well-being, not least since the judgement that one's testimonies have been disregarded tends to undermine one's ability and willingness to engage in further interpersonal exchanges. This suggests that the style of patient–clinician interactions should involve respect for the patient perspective, even in cases where the epistemic credentials of that perspective are less than impeccable. Those interactions do not, after all, consist of an atomized series of decontextualized propositions. Instead, those interactions involve a dialogical relationship, in which cumulative experiences—of silencing, say, or trusting—affect the structure and content of later stages of the interaction.[8]

An ill person can suffer testimonial injustice in one of several ascending ways. At the simplest level, ill people's testimonies can simply be ignored by healthcare professionals, perhaps being heard but neither acknowledged nor considered. Or those testimonies may be heard but excluded from epistemic consideration; so a genuinely sympathetic clinician might listen to her patients' testimonies but fail to see that those testimonies may be worthy of epistemic consideration in virtue of being useful or informative, say. Patient testimonies may be heard and

---

[7] An example of miscommunication causing the patient to stop treatment is engagingly described here: <https://thefilthycomma.wordpress.com/2012/01/17/busting-a-gut/> (accessed 3 November 2015).

[8] More generally, we are not denying the existence of epistemic asymmetries, but call for discussion on how these should be managed in cases where they exist and clinicians are aware of them. We would agree that a teacher who dismisses students or talks down to them on the basis of their epistemic inferiority is not a good teacher. Dialogical openness is compatible with epistemic asymmetry and can serve to address the imbalance. Distinguishing the stance adopted by a clinician from the perceived and actual epistemic status of claims made by clinician and patient might be useful. We thank an anonymous referee for emphasizing this point.

acknowledged, but judged to be irrelevant or insufficiently articulate, and so once again excluded from epistemic consideration. A patient may lack the language and concepts to express their situation, perhaps resorting to clichés that clinicians judge to be too coarse or idiomatic to be of use, or feeling that they ought to employ formal medical vocabularies in which they know themselves to be inexpert. Or those testimonies may be acknowledged but subordinated to the authority of health professionals, so what the patient has to say is heard, but coupled to the conviction that it is the view of health professionals (and in particular physicians) that is authoritative or primary. The ill person may be judged to be a useful informant but not a participant in the collective practice of interpreting and understanding their medical situation.

These indicate some of the strategies, implicit or explicit, by which the testimonies of ill persons can be excluded or downgraded in a way that secures the charge of epistemic injustice. Importantly, we do not suggest that these strategies are systematically employed consciously or deliberately and certainly not with malice (although they may be). Rather, we are pointing to a set of practices and behaviours based on presuppositions and assumptions that are rarely reflected on and are largely tacit (compare Fricker 2007, p. 38).

To demonstrate this point here are two examples of inappropriate credibility assigned to patient testimonies. The first example is that of Myalgic Encephalomyelitis/Chronic Fatigue Syndrome (ME/CFS) patients, whose disorder is not recognized by many physicians or considered a psychiatric (rather than somatic) illness. In the 2011 documentary *Voices from the Shadows*, directors Josh Biggs and Natalie Boulton interview ME/CFS patients, who report distinctively somatic symptoms such as pain, sensitivity to light and noise, and fatigue. However, because the medical and nosological status of ME/CFS is contested, these reports are disbelieved or subsumed under a different interpretation (e.g. that the patients suffer from abuse or that they have a psychiatric illness).

Here are a few examples. One patient says: 'my suffering was belittled' (2:25). Another comments on the high degree of disbelief in the reality of ME/CFS as a *bona fide* disease (8:00). And a family reports that 'belief turns to disbelief when tests come back normal' and describes how 'professionals turn against the family' suspecting that the family is harming the child suffering from the condition (34:39). In extreme cases, children with ME/CFS are removed from their family, care proceedings initiated, and children placed in psychiatric units or in foster care. These practices stem from a deep disbelief in the reality of ME/CFS as a somatic medical disease. Given this context, it is not surprising that patients' and families' testimonies are accorded little credibility and their interpretation (that the child has ME/CFS) is rejected in favour of another interpretation (that the child suffers abuse or has a psychiatric disorder). In other cases, clinicians and patients will agree on the diagnosis, but the clinicians consider the conditions as a mental disorder, while the patients claim it is somatic.

Contrast this with Munchausen syndrome (factitious disease), in which people produce or feign disease (e.g. by taking laxatives or wounding themselves) for complex psychological reasons.[9] In these cases, patients take up clinicians' time and are given considerable medical attention although the symptoms are feigned or made up (Savino and Fordtran 2006). Against the backdrop assumption that no one would seek medical help unless they were ill, patient reports of symptoms are normally believed. Only after repeated visits to the doctor, medical facts that contradict the patient's story, or the absence of supporting medical documentation, do health professionals revise the level of credibility assigned to such patients' testimonies.

These two examples illustrate the epistemic discretion exercised by health professionals when listening to patient testimonies and the two ways in which credibility assignment can be faulty: too high or too low. However, this usually tacit credibility assessment taking place in patient–clinician encounters is not explicitly addressed in medical education and training and indicates the need for attention to this phenomenon.

One may ask in particular whether the status of 'patient experts' may prevent epistemic injustice when patients with a chronic condition know their own condition well and have a good grasp of the medical facts and the state of scientific knowledge about it. That is certainly an important step towards acknowledging informal expertise. However, to assign someone a status of epistemic authority ('patient expert') is in itself

---

[9] Baron Münchausen (1720–97) was a German war hero who travelled around Germany describing his military adventures. There is no evidence that he feigned disease or duped people into caring for him. As Feldman (2004) notes, Rudolph Erich Raspe appropriated this name for the title of a 1785 pamphlet of fanciful and patently false tales, *Baron Munchausen's Narrative of His Marvelous Travels and Campaigns in Russia*.

insufficient unless one also adjusts the wider structure of epistemic norms and practices to 'build in' those new authorities. For example, rheumatic patients who are members of a physiotherapy group might be classed as 'patient experts', but are still not invited to participate in (for instance) the physiotherapy committee, or consulted about changes to the hydrotherapy service. Expertise can be misconstrued if thought of individually; established roles for patient experts are also needed within the wider set of structures and practices of the relevant context.

What would testimonial justice look like? In testimonial justice the testimonies of ill persons are recognized, sought out, included within epistemic consideration, judged to be relevant and articulate (where they are), and at least in certain respects, judged as epistemically authoritative. The testimonially just clinician is 'alert to the possibility that her relative unintelligibility to him is a function of a collective hermeneutical impoverishment, and he adjusts or suspends his credibility judgement accordingly' (Fricker 2007, p. 7). So a testimonially just clinician, confronted with an emotionally charged patient testimony whose medical value she cannot discern might think: 'The fact that I don't understand you isn't your fault, but mine; even your best efforts to make yourself understood are failing, not because of their inarticulacy, but because I am untrained in the appreciation of the sort of articulacy you are using, and this hermeneutical context does not provide me with those resources.'

Similar considerations apply to cases of hermeneutical injustice; hence our treatment of this will be brief. Hermeneutical injustice occurs when someone's testimony is not squarely disbelieved but a conceptual impoverishment in a particular culture prevents that person from being able to clearly articulate their testimony. This generates what Fricker calls 'a gap in collective hermeneutical resources' (2007, p. 7). Fricker gives as an example the case of sexual harassment in a culture in which the concept itself does not exist. How would one go about making the relevant assertions if one lacks the concepts to do so? Importantly, hermeneutical injustice needs to be cashed out in contextual terms, e.g. showing what sorts of practices or social norms or institutional structures generate it. Articulating this injustice is the task of this section.

In the case of illness, the interpretations that ill persons make of their own experiences may simply be ignored or not sought out, or implicitly excluded through the establishment of a culture in which patient views are not respected or included within policy, thereby withdrawing

incentives for ill persons to offer their interpretations. Or patient interpretations may be heard and considered but judged to be irrelevant or
insufficiently articulate, perhaps because they are too bound up with
'subjective' concerns and anxieties or because the practice of taking them
seriously is not recognized as being of value.

Or those interpretations may be reductively seen as another source of
information that can be assessed or utilized by clinicians as if those
hermeneutical offerings were simply data and so not treated as being
an epistemically distinctive form of knowledge. Later in the chapter
we appeal to phenomenology to indicate how a tendency to elide
first-person accounts with third-person reports can have deleterious
consequences. However, it is important to note that as well as being
harmful, it is an epistemic error in itself to collapse important distinc-
tions between first- and third-person reports because it deprives us of
sufficiently nuanced epistemic resources.

An ill person experiences hermeneutical justice, by contrast, when the
interpretations of ill persons are recognized, sought out, included within
epistemic consideration, judged to be relevant and articulate (where they
are), and at least in certain aspects, judged as epistemically authoritative.
An example of epistemic justice, incorporating both testimonial and
hermeneutical justice, is the case of Kingston General Hospital (KGH)
in Ontario, Canada. Following financial failure and high rates of patient
complaints, KGH was redesigned using patient input at each step. The
hospital has a Patient and Family Advisory Council and patient experience advisors are members of key hospital committees.[10] Every decision
made at KGH must have a patient included in its consultation or provide
a reason why patients were not consulted; patient–health professionals'
codesign is fundamental to the hospital's practice.[11]

8.2 Examples of Epistemic Injustice in Illness

The foregoing accounts are schematic and should not be interpreted as
a rigidly categorical description of the 'stages' of epistemic injustice.
Epistemic injustice must not be conceived in terms of abstract

---

[10] See <http://www.kgh.on.ca/en/aboutkgh/Patient%20and%20Family%20Advisory%
20Council/Pages/default.aspx> (accessed 29 October 2015).
[11] Leslie Thomson, KGH Chief Executive, talk at King's Fund, 8 November 2012.

epistemological analysis alone, says Fricker, for it must be sensitive to a 'socially situated account' which recognizes that human beings *qua* epistemic agents are recognized as 'operating as social types who stand in relations of power to one another' (Fricker 2007, p. 3). Therefore the epistemic injustice that is experienced by ill persons must be sensitive to their social situation, comparative credibility, and so on, including factors such as intergenerational variation in doctors' attitudes towards their patients.

We also need to identify different degrees of injustice, to help pin certain sorts of injustice to certain behaviour, to make easier the task of correlating forms of injustice to specific policies and practices. Perhaps more conservative and authoritarian doctors might simply ignore patient testimonies, whereas others do not. It also may be the case that health professionals would welcome patient testimonies, but the acceptable formats for collecting such information do not suit the kind of testimony patients wish to share, e.g. using yes/no questions where nuance and context are essential; giving limited space to describe an event or experience; asking only about specific aspects of healthcare provision; or asking patients to fill in such questionnaires in public spaces with little time and privacy.

Here are some examples of testimonial and hermeneutic injustice:[12]

I asked a professor whether being exposed to reduced oxygen levels long-term, the way I am, would have any detrimental effects on cognitive function e.g. would that explain why my memory had rapidly become much worse? He just laughed off my genuine and serious concern by saying he had the same problem and sometimes couldn't even remember his wife's name. I never did get a proper reply to that question.

I don't mention problems because though they are real for me, they're minor in the grand scheme of things.[13]

I had an abnormal cervical smear, so was sent to the large city teaching hospital for a coloscopy. I changed into the usual ties-up-the-back gown, with the usual vital ties missing, and then went through for the examination. It's a bit uncomfy but I was ok. Lots of big sighs from the consultant with his head between my legs. Then off he goes, leaving the room. I'm told to follow. So I arrive, naked under a gown which doesn't do up, slightly damp between the legs

---

[12] These examples are taken from responses to a query we posted on a patient mailing list in 2012.

[13] Self-censoring is another form of epistemic injustice, in which the negative stereotyping is internalized by the patient herself, leading her to downgrade her own testimony.

and a bit stressed as I have to sit down and I'm worried about leaving a wet patch. He goes on to tell me I need an operation. I hear blah-blah-blah as I'm perching and panicky. And it's very difficult to think without your pants on. I said nothing.[14]

A disabled person complains that friends always ask 'What did the doctor say?' without either the follow-up question 'And what do you think about what she said?', or by implicitly treating the ill person as a 'testifier' (a source of information) and the doctor as the 'interpreter and actor' (who acts on the information) (personal communication). Leontiou, a mother of a disabled child, concurs when she writes:

What I find most striking is that, when I reflect on the good interventions that I have brought to my son, most have been recommendations from other mothers. Doctors don't offer many ideas for navigating the world of disability. Yet, I am repeatedly asked, 'What do the doctors say?' I don't know exactly how to answer this question. Rather, I'm interested in examining how asking this question places the doctor in a central position and gives the impression that the doctor is the only one who knows. I have never been asked, 'What do other parents who are in your circumstance say?' (Leontiou 2010, p. 2).

More extreme historical examples are the following. The first is the case of curare, a poison that causes paralysis that was used as a general anaesthetic for major surgery in the 1940s under the misapprehension that curare was a general anaesthetic. As Daniel Dennett writes:

The patients were, of course, quiet under the knife [. . .] but when the effects of the curare wore off, complained bitterly of having been completely conscious and in excruciating pain. The doctors did not believe them. (The fact that most of the patients were infants and small children may explain this credibility gap). Eventually a doctor bravely committed to an elaborate test under curare and his detailed confirmation of his subjects' reports was believed by his colleagues (1981, p. 209).

Another example also involves the use of anaesthetic. David Wootton (2007) describes how nitrous oxide was discovered and its analgesic properties noted in 1795, but only put into use as anaesthetic in 1846. He writes:

[Y]ou need to imagine what it was like to become so accustomed to the screams of patients that they seemed perfectly natural and normal; so accustomed to them

---

[14] In this instance the patient is put at a situation that is disadvantageous to her *qua* epistemic agent: she is stressed, worried, and humiliated about being undressed. She cannot be expected to communicate well under such circumstances.

that you could read with interest about nitrous oxide, could go to a fairground and try it out, and never imagine that it might have practical applications (2007, pp. 22–3).

Let us offer one final example, that of a psychiatric patient who also has a physical disorder. Here is Elyn Saks' (2007) account of her brain haemorrhage:

> Quickly, they bundled me into the car and took me to the emergency room. Where a completely predictable disaster happened: the ER discovered I had a psychiatric history. And that was the end of any further diagnostic work. [...] Poor Maria was literally jumping up and down trying to tell anyone who'd listen that she had seen me psychotic before and that this was different. But her testimony didn't help—I was a mental patient. The ER sent me home (2007, pp. 232–3).

Saks also recounts the story of a psychiatric patient who went for weeks with a broken back, because none of the medical staff took his pain seriously (2007, p. 232). These last examples may seem extreme; we present them in order to demonstrate that epistemic injustice in the case of illness can have devastating effects and can range from the subtle and hard–to–detect bias to brutal rejection of clear evidence of suffering.

Such iterated experiences give rise to the self-propagating nature of such acts of exclusion: the patient's testimonials are ignored or down-played, which upsets the person offering testimony and interpretation. This on its own is wrong and gives rise to the common complaint that 'the doctor doesn't listen to me'. But it also affects future epistemic offers, so testimonials may become infused with self-doubt and emotionally charged, therefore confirming the doctors' perception of that patient as a poor testifier, leading to a vicious circle of damaging communication.

8.3 Epistemic Privilege

Another side to this discussion of epistemic injustice is the epistemic privilege accorded to health professionals. This privilege is accorded by virtue of their training, expertise, or third-person psychology, such that they occupy the epistemically privileged role of assessing which testi-monies and interpretations to act upon, as well as deciding what sorts of testimonies to receive, from whom, what form they can take, and so on. In this section we will discuss this more elusive kind of epistemic injustice and relate it to the difference between patient and clinician attitudes to illness.

We claim that the structures and discourses of contemporary health-care practice might encourage epistemic injustice because they privilege certain styles of articulating testimonies, forms of evidence, ways of presenting and sharing knowledge, and so on. We claim that modern healthcare practices privilege impersonal third-person reports and empir-ical data over personal anecdote and pathographic testimonies, in a way that structurally disables certain testimonial and hermeneutical activities. Different kinds of epistemic injustice can occur separately or could be mutually reinforcing; where all three are present and active one has what Fricker calls 'persistent systematic epistemic injustice' (2007, p. 58).

Many health professionals might like to spend more time and energy taking seriously patient testimonies, but the pressures—of time, resources, task-based organizational processes, etc.—that they operate under make this very difficult. Many health professionals may be unwillingly epistem-ically unjust and would like this aspect of their work to be different. Structural and hierarchical features of the healthcare system are the cause of the epistemic injustice, rather than any individual's intention.

Both health professionals and ill persons are epistemically privileged for different reasons. But only the health professionals' privileged epistemic status 'really matters' when it comes to healthcare practice and policy. The knowledge of patients is usually confined to the private realm and is not readily incorporated into decision-making, intervention design, and policy (Wainwright and Macnaughton 2013). In recent years the terms 'patient centred-care', 'patient expert', and 'patient experience' have become common in policy documents and mission statements, and we hope that with time these translate into actual improvement to healthcare provision.[15] But healthcare failures are still abundant and reflect the institutionalizing and morally paralyzing force of some current healthcare provision arrangements.[16]

---

[15] See for example the work of organizations such as the King's Fund and the Point of Care Foundation and, for the UK National Health Service, Coulter and Ellins (2006), Greener (2009), and McIver (2011).

[16] A vivid and tragic example is the series of systematic failures which led to the death of hundreds of patients, uncovered by the Mid Staffordshire NHS Foundation Trust Public

There are several reasons for the epistemic privileging of health professionals, and in particular of physicians. First, on the medical view the goal of medicine is to repair physiological mechanisms. The third-person view dominates this model and has no obvious room for first-person testimonies. Second, in a performance-based, target-driven culture patient input has little place. Third, in a large-scale healthcare system in which performance is judged quantitatively, qualitative statements are difficult to utilize. Fourth, patient views are often seen as anecdotal and context-dependent and therefore lacking any long-term utility. Finally, patient views can be as numerous as patients and therefore it is unclear whose views should be acted on.

The knowledge each group might bear is different. Patients have the knowledge of how a particular condition feels, how it impacts on their life, and changes their way of being in complex and subtle ways (Carel 2010, 2013a). Only they can say whether a certain treatment causes pain, or how well they feel. Clinicians have the scientific, medical, and clinical knowledge. Of course the two domains of knowledge do not belong exclusively to one epistemic group. For example, a clinician with an extensive experience of treating a particular disease may have excellent knowledge of the limitations on daily living it may impose on patients although she has no first-person experience of the disease. Similarly, the 'patient expert' may develop a deep understanding of their condition and its causes, as well as be an authority on treatments, trials, and research in the field. Moreover, the two groups may work collaboratively to integrate and promote both kinds of knowledge (Rosenbaum 2012). These two domains of knowledge are different also in how such knowledge is gained (Kidd 2013).

What is taking place in the epistemic domain with respect to these two groups and these two kinds of knowledge needs to be related to broader issues concerning the relationships of priority and power, as well as reciprocity, between different epistemically privileged groups. Thus for example, patients are expected to be told what to do by doctors, but doctors are not expected, bar rare cases, to be told what to do by patients. There is an asymmetry in the relationship owing to an implicit hierarchy assigning the health professional (and especially physicians) a high

Inquiry, led by Sir Robert Francis in the UK. See <http://www.midstaffspublicinquiry.com/report> (accessed 29 October 2015).

epistemic status that is linked to a professional and widely acknowledged social position.

The term 'epistemic privilege' has three related components. A person or social type ('consultant') may be epistemically privileged because they have the authority to establish, and where necessary to enforce, the standards and norms for epistemic exchange in a given community. For instance, the medical community is epistemically privileged because it can define and characterize medical concepts (such as 'health' and 'disease') and so sets the terms for authoritative debates about health and healing.[17] Although this does not prevent ill persons from having parallel debates, it ensures that their debates are not considered authoritative.

A person or social type may be epistemically privileged because they occupy an authoritative procedural role in epistemic exchanges, for instance by acting as gatekeepers, controlling which persons and groups are included and what degree of credibility and authority they are assigned, and acting to enforce discipline within the epistemic community. For instance, a hospital review committee is epistemically privileged because it has the authority to decide how to populate the committee, who are permanent and who are invited members, which persons have secondary status (such as 'observer'), who acts as chair, what the agenda for debate is, and so on. A person or social type may be epistemically privileged, finally, if they have what one might call power of decision, that is, if it is their privilege to decide when an issue is settled, when enough evidence has been presented, when a particular issue has been given sufficient time and attention, and so on.

These three forms of epistemic privilege are likely to arise together and be mutually reinforcing. Consider a hypothetical case in which a group of patients with chronic rheumatic disease are invited to sit on a committee reviewing the physiotherapy provision available to them. Those patients might suffer epistemic injustice in this case because they lack epistemic privilege in the three ways articulated above. First, they are denied any opportunity to determine whether or not the definition of the concept of 'health' being used is appropriate or consonant with their experiences; for instance if health is defined in terms of their performance of

---

[17] The fact that the medical community has these forms of social and epistemic power does not, of course, entail that they always exercise that power in a robustly procedural manner (see Kidd 2013).

physiotherapy exercises rather than their capacity to perform everyday tasks like driving. Second, their epistemic authority is minimized because their status is that of 'invited observers' who can be consulted but who have no substantive critical powers. Third, those rheumatic patients lack any decisive role in the review committee because they have no power of vote and so cannot enforce their insistence that certain issues be discussed more thoroughly.

We do not wish for this debate to sound one-sided or blind to the considerable merit—epistemic and otherwise—of medical training and practice. We do not aim to attack any specific epistemic group, but point to the current arrangements that give rise to epistemic injustice at a considerable cost to patients and possibly to health professionals who are constrained by these very practices. We do not suppose that culpability for epistemic injustice should be placed at the feet of health practitioners; the attitudes and actions of those practitioners will be shaped by particular models of the patient–clinician relationship which they were trained in or are required to operate with. It may be that certain models are, however inadvertently, more liable to generate receptive conditions for epistemic injustice than others.

Emanuel and Emanuel (1992) identify four models of the patient–clinician relationship; each incorporates implicit epistemic presuppositions. For instance, a 'paternalistic model' establishes a strict epistemic asymmetry in which the doctor authoritatively informs the patient of what is best for them, whereas a 'deliberative model' encourages a dialogical exchange between patient and clinician that, at least in principle, affords greater epistemic autonomy to the patient. Indeed, Emanuel and Emanuel (1992, p. 2226) conclude their paper by noting that the models may incorporate 'defective conceptions' of epistemically charged concepts such as 'patient autonomy', and could be criticized on those grounds.[18]

To this end we claim that it is useful to distinguish between warranted and unwarranted epistemic privilege, e.g. healthcare professionals warrant epistemic privilege in their interpretation of a CT scan but not in deciding where a patient should die (e.g. in hospital or at home). We fully acknowledge that certain persons and professions are epistemically privileged,

---

[18] Indeed, the issue of culpability for epistemic injustice is complex, for instance because it varies according to whether the particular form of injustice is *agential* or *structural*. See the exchange between Riggs (2012) and Coady (2012).

at least in certain cases. But there may be cases where the epistemic privilege of health professionals is unwarranted; for instance, if the assignment of epistemic privilege is grounded in the presumptive judgement that there are no other plausible candidates for privileged epistemic status in the context of certain forms of patient care. For example, many religious persons who experience depression consult persons they recognize as having spiritual authority—priests, say—as well as psychiatrists and other mental health professionals (compare Scrutton forthcoming; Kidd forthcoming). Some argue that doctors are exempt from certain charges of epistemic error—e.g. their being hoodwinked by the biases in the academic literature concerning the efficacy of drugs—because they are simply too busy to perform the complex processes of survey and analysis (compare Goldacre 2012).

8.4 Addressing Epistemic Injustice: A Phenomenological Patient Toolkit

So far we have examined epistemic injustice in illness. We argued that illness might give rise to testimonial injustice, when patient claims are ignored or rejected, and to hermeneutical injustice, when patients do not have the concepts with which to articulate their illness experiences. In this section we claim that not only is this epistemic injustice damaging, but that certain experiences of illness can afford epistemic privileges to the ill person that are not otherwise available to (and perhaps not fully shareable with) persons not ill.

This idea has important ethical implications: it calls us to take seriously the standpoint of ill persons, militates against paternalism, and acknowledges the essential role that ill persons should play in the formulation and implementation of healthcare policy. We suggest that phenomenology can give us some essential tools for overcoming epistemic injustice and can therefore contribute to explicating the experience of illness. We further suggest that phenomenology is intimately connected to issues of epistemic justice, since two core phenomenological ambitions are to (i) identify and articulate the tacit structures that underpin one's experience and engagement with the world, and (ii) to provide a means of taking seriously the (often radically different) experiences of others (compare Ratcliffe 2012b).

The idea behind the toolkit is that philosophical concepts can aid patients who are trying to make sense of their illness, as well as health professionals caring for them. Patients may have ethical, existential, or metaphysical concerns that have a distinctive philosophical content, but are not conceptualized as such. For example, those nearing the end of life may reflect on its shortness and wonder whether they have lived a good life. Developing and articulating such reflection is a distinctively hermeneutical activity, albeit of a unique sort. These reflections are foisted upon the ill individual by their illness; the context in which this philosophical activity takes places is difficult and physically and emotionally taxing; and such reflection stems from concrete and idiosyncratic concerns rather than from engagement with abstract questions.

In order to enable this hermeneutical activity, a toolkit has been developed that helps patients understand the impact illness may have on their life (Carel 2012). It provides philosophical concepts through which the impact of illness, and of caring for the ill, may be interpreted and conceptualized. These concepts are taken from phenomenology and utilized in order to provide an account of the total nature of illness (Carel 2013a). The practice may help ameliorate epistemic injustice by giving patients the ability to interpret and articulate their illness experiences, thus addressing the hermeneutical gap Fricker reveals.

This toolkit is not intended to replace medical knowledge but to supplement it. It answers a need identified by patients as well as health professionals to better understand the illness experience of individuals and groups with particular conditions. For example, a GP pointed out the need for narrative humility among clinicians, and the personal growth that listening to patients in a holistic way would bring (personal communication). There is a knowledge that arises from having a particular illness experience that should not be dismissed as idiosyncratic or too emotive, as it crucially interlaces with and illuminates the medical facts. We propose that adding first-person accounts of the illness experience to the overall medical picture may make diagnostic, therapeutic, and empathetic contributions to healthcare provision. Incorporating patients' insights can help with the diagnosis and treatment of illness and also enrich clinicians' empathy as well as guard against alienation.

This involves the transition from the 'informational perspective', which sees the speaker as a 'potential recipient or source of information' to the 'participant perspective', in which we see the quest for knowledge as a shared enterprise and the patient speaker as 'competent to carry out some particular activity that has a fundamental role in carrying out inquiries' (Hookway 2010, pp. 156–7). As Hookway suggests, 'there could be a form of injustice related to assertion and testimony that consisted, not in a silencing refusal to take the testimony to be true or expressing knowledge, but in a refusal to take seriously the ability of the agent to provide information that is relevant in the current context' (2010, p. 158; for a detailed analysis of this point see Kidd and Carel 2016). It seems to us that it is frequently the ill person's ability to offer relevant assertions that is being questioned, rather than their ability to make assertions at all.[19] The patient in such cases is 'recognized as unable to participate in activities whose content is intrinsically epistemic' (2010, p. 159). A forum in which patients can gain epistemic confidence and discern their experiences of illness might help tackle epistemic injustice by supporting patients' transition from informational to participant perspective.

The toolkit is a patient resource, but it is also aimed at training clinicians. If clinicians are trained in this way and, consequently, become more open to patients' experiences and better able to interpret them, this would be yet another way to address the hermeneutical gap discussed above. Ideally, the shared use of the toolkit would help promote epistemic symmetry, in which the patient is more empowered and articulate, and the clinician more aware of the richness and value of first-person accounts.

Here is an outline of the toolkit (see also Carel 2012). It provides a flexible individual tool which patients and clinicians can use to develop their understanding of their illness experiences. It includes three steps: bracketing the natural attitude, thematizing illness, and reviewing the ill person's being in the world. We discuss each in turn.

Serious illness removes our conventional understandings and expectations and is thus an opportunity to examine choices, routines, and values. Merleau-Ponty writes that reflection 'slackens the intentional threads which attach us to the world and thus brings them to our notice' (2012, p. xiii). Illness enables such withdrawal, because it imposes a

---

[19] The case of some mental disorders (e.g. psychosis) would be different. In these cases the patient may be considered altogether irrational and largely unable to make true assertions. (see Crichton, Carel and Kidd, in press)

re-evaluation on the ill person (see Chapter 9). A phenomenological approach to the experience of illness requires a suspension of a 'natural attitude' of implicitly accepting the background sense of belonging to a world and various interpretive dogmas along with it.

Bracketing the natural attitude is a withdrawal from the ordinarily implicit commitment to the reality of the world (Ratcliffe 2008a, p. 4). As Husserl says, this is not a sceptical or idealist position. Rather, this 'inhibiting' or 'putting out of play' of the natural attitude exposes 'the universe of phenomena in the phenomenological sense' (Husserl 1988, p. 20). This suspension of the everyday routine understanding of illness allows its under-theorized aspects to become an object of inquiry because it enables us to shift attention from the disease entity to the way in which it is given and its modes of appearance to us.

The first step of the toolkit, bracketing the natural attitude toward illness, suspends the belief in the reality of an objective disease entity. Shifting the focus away from the disease entity and toward the experience of it can disclose new features of this experience, to both patients and clinicians. We usually take the disease entity for granted and posit it as the source of the illness experience. This approach is particularly common within healthcare professions. But in fact, for the ill person the illness experience comes before the objective disease entity (Toombs 1987). Once the belief in the objective disease entity is bracketed and we are distanced from our usual way of experiencing, we can explore how illness appears to the ill person, and what essential features it might have.

The second step in the toolkit is thematizing illness. 'Thematizing' refers to the act of attending to a phenomenon in a way that makes particular aspects of it explicit (Toombs 1987, p. 222). A theme for a particular consciousness is that upon which it focuses its attention. But this does not simply denote the intentional object. It also takes into account the kind of attentional focus given to an entity. Thematizing may include attending to the cognitive, emotive, moral, or aesthetic aspects of a phenomenon. A patient may thematize her illness as a central feature of her life, attending to her symptoms as pervasive, while the clinician may thematize the illness as a 'case of cancer', attending to symptoms as diagnostic clues (p. 222).

The understanding that illness is not an objective entity and the exercise of thematizing may help both patients and clinicians because it enables moving away from prescriptive pronouncements toward a more tentative, descriptive mode. Thematizing can be used to bring out the multiple perspectives on one's illness that patient, family, health professionals, and others may have, as each will thematize illness differently.

A mixed workshop, with both patients and health professionals, would be ideal for carrying out this thematizing. The patient may thematize her illness primarily emotively, while a health professional will thematize it cognitively. A family member may thematize illness as an experience of empathy. Exploring the different thematic centres illness may have can illuminate its multiple ways of appearing.

The third step of the toolkit is to take the new understanding of illness emerging from these two steps, and examine how it changes the ill person's being in the world. Being in the world includes the biological entity, the person, and her environment and meaningful connections (Heidegger 1962). The toolkit uses being in the world to capture the pervasive effects illness may have on one's sense of place, on one's interactions with the environment and with other people, on meanings and norms, and on the nexus of entities, habits, knowledge, and other people that makes up one's world. The third step enables participants to move away from a narrow understanding of illness as a biological process, towards a thick account of illness as a new way of being in the world.

The toolkit uses Heidegger's notion of 'being in the world' to capture the pervasive effects illness may have on one's sense of place, interactions with the environment and with other people, meanings and norms, and the nexus of entities, habits, knowledge, and other people that makes up one's world. This term enables us to elaborate richly and comprehensively on the impact of illness. By moving away from a narrow understanding of illness as a biological process, a thick account of illness as a new way of being in the world can be developed by patients. Because illness turns from being an external intrusion to being a form of existence, the notion of being in the world is particularly appropriate. It helps understand the pervasive impact illness may have on all life domains, which are seen as interconnected.

One way in which the toolkit can be delivered is as a workshop for patients and health professionals. It could use visual and sensual samples, as well as texts and philosophical ideas, to trigger discussion and reflection. The evocative force of images and sounds will enable participants to explore possibly unnamed emotions and experiences. The phenomenological dimension of the workshop is amplified by this use of varied media, which will appeal to the experiential and perceptual, rather than restrict exploration to already formulated ideas. The workshop will reflect the insight that speaking about the experience of illness is not just a means for expressing already formed thoughts but that speech brings thought into being by allowing it to form (Merleau-Ponty 2012; Gallagher 2005). The workshop will also use non-linguistic means for self-description and self-reflection, for example, asking participants to create a collage of their situation or choose a song that describes how they feel.

Such a workshop would begin by presenting the three steps of the toolkit, explained within the context of illness. Each step will be used to trigger discussion, combined with small group exercises. The three steps will then be used to analyse the experience of illness, using categories such as space and time, lost abilities, and adaptability. The exploration will then move on to use a variety of media (film clips, texts, images, and music), which will be tailored to the specific context of the group.

The toolkit will enable the expression of unique personal experiences rather than pushing patients to adapt their experiences to medical or cultural expectations. The small-group structure of the workshop and the fact that participants all suffer from an illness, or aim to care for ill persons, provide a safe environment that will allow participants to share the idiosyncrasies of their experiences with no pressure for these to fit into a pre-given mould. Of its very nature, the illness experiences of different patients will contain conflicting understandings. Such conflicting understandings do not need to be resolved because no single understanding is offered by phenomenology.

The toolkit has been shared with patient groups as well as with a group of GPs in a consultative process. Initial reactions to it demonstrate the feasibility and helpfulness of such a tool.[20] For example, the GP group consulted suggested the toolkit would be of particular use with elderly patients and those suffering from depression. Patients suggested that the toolkit would help them by empowering them to 'speak their mind' and offering an opportunity for reflection rather than self-pity. Such a toolkit would allow a space for grieving and would enable patients to take

---

[20] These comments were collected during three consultative sessions: one with a group of GPs took place on 14 June 2012 and two patient group sessions took place on 14 and 21 September 2012, all in Bristol, UK.

responsibility for their understanding of illness by enhancing their self-knowledge. We are not suggesting that the toolkit is a sole way of addressing epistemic injustice; far from it. We suggest it as an example of one possible practical measure that, in conjunction with other measures, may begin to address the problems outlined in this chapter.

8.5 Conclusion

We conclude that the phenomenon of epistemic injustice identified by Fricker is likely to be much more widespread within healthcare contexts than it would seem given the great emphasis on patient experience, autonomy, and values within healthcare literature. Fricker herself comments: 'I believe that there are areas where injustice is normal and that the only way to reveal what is involved in epistemic injustice (indeed, even to see that there is such a thing as epistemic injustice) is by looking at the negative space that is epistemic injustice' (2007, p. viii).

It would be interesting to consider the possibility of epistemic injustices that might arise in other contexts—for instance, between social workers and 'clients', prison inmates and wardens, and police officers and suspects. Such comparative analyses might expose some similarities in styles of epistemic exclusion that characterize particularly asymmetrical power relations. We hope to have begun this task by discussing the case of epistemic injustice in ill health and in the exchanges between patients and health professionals.

Fricker writes:

[P]rejudice tends to go most unchecked when it operates by way of stereotypical images held in the collective social imagination, since images can operate beneath the radar of our ordinary doxastic self-scrutiny, sometimes even despite beliefs to the contrary [. . .] our everyday moral discourse lacks a well-established understanding of the wrong that is done to someone when they are treated in this way (2007, p. 40).

It is our hope that this chapter will contribute to the effort of lifting stereotypes and biases about ill people and about illness from the unconscious collective imagination to the conscious level and therefore make it available to the careful scrutiny it deserves.

9

The Philosophical
Role of Illness

For what is it to be ill? Is it that you are near the severance of the
soul and the body? (Epictetus, *Discourses*)

We have now come full circle from the Introduction.[1] We started out by
asking how phenomenology can be used to illuminate illness. The earlier
chapters of this book examined how this illumination may take place; i.e.
how philosophy can be used instructively to understand the experience
of illness and its deep structures. Phenomenology was chosen as a useful
method for describing lived experience, and in particular the lived
experience of illness. We examined how illness forces the ill person to
modify her relationship to her environment and to other people, as well
as her attitudes towards life, death, and value. The book then presented the
notion of bodily doubt, and looked at a particular example of a phenomenology of illness, namely, the experience of pathological breathlessness.
We then asked whether illness negatively impacts on happiness, answer-
ing that it does not radically diminish well-being. We then turned to
Heidegger's notion of death and structure of human existence as finite
temporality. And finally, We suggested that the analysis of interactions
within healthcare settings through the notion of epistemic injustice can
reveal the credibility deficits to which ill people are more vulnerable.

In this final chapter I would like to turn to the other side of the
bilateral relationship between illness and philosophy. Not only does
philosophy illuminate illness, but illness, I suggest here, can illuminate

---

[1] This chapter was first published as 'The Philosophical Role of Illness' (2014), *Metaphilosophy* 45(1): 20–40. Minor changes have been made to the paper to adapt it to this chapter.

philosophy. This chapter examines the philosophical role of illness, claiming that the role is important but overlooked, and providing examples of its contribution to philosophical work.[2]

It is worth noting at the outset that many areas of human life that are seen as central to it, such as family, love, health and illness, and well-being, are poorly reflected in the history of philosophy, and especially so in contemporary academic philosophy. Issues that are existentially important to most people, such as ageing, parenting, and the environment, have traditionally received little philosophical attention. Even when those topics are discussed, they are often discussed narrowly, e.g. in the context of discussing the ethics of parenting, rather than within the context of a systematic exploration of the place and significance of relationships in human life.

Illness is a case in point for such marginalization. The philosophy of medicine (or the philosophy of health and illness) is still not considered a mainstream philosophical topic, but an optional area of specialization. One might retort that this criticism implies that philosophy ought to 'cover' human life, but that this demand is unjustified. I do not wish to make this claim but to point out the oddness in the lack of philosophical discussion of such core areas of life. The reply that perhaps there is nothing of philosophical interest in those areas is question-begging; surely it is the task of philosophical exploration of this sort to find out what there is to say philosophically about these areas and how they might link up with debates in ethics, political philosophy, or metaphysics. Merely relegating them to the pile labelled 'private', 'subjective', or 'anecdotal' is deeply unsatisfying.

A vivid analogy can be drawn with feminist philosophy that has gradually come to tackle topics such as domestic violence, sexual harassment, childbirth, and sexuality—which were initially considered to be of little philosophical interest. I believe that illness, ageing, vulnerability, and childhood are yet to follow suit.[3] In addition, the philosophical

---

[2] In this chapter I continue to use the term 'illness' to denote serious, chronic, or life-threatening illness, rather than common and transient illnesses, such as flu. However, less serious conditions can also be philosophically important, as they disclose more minor interruptions to the flow of experience. Sartre (2003) gives the example of a headache disrupting reading, which is discussed in Chapter 2.

[3] For a recent collection on the philosophy of ageing see Cottingham (2012). On childhood in the history of philosophy see Krupp (2009).

neglect of illness stands in stark contrast to the enormous interest in illness in other academic disciplines, e.g. sociology, anthropology, law, history, and bioethics, to name a few, and as can be seen in the recent explosion in medical humanities work. It is also surprising given the great interest in illness and healthcare among the general public (e.g. the enormous popularity of pathographies, online fora and discussions on health, and popular writing in medical history, e.g. disease 'biographies').[4]

In contrast with modern-day philosophy, illness has been a theme in the history of philosophy, in particular in relation to its moral, existential, and spiritual value. For example, the first-century AD Stoic philosophers Epictetus (2004) and Seneca (2004), second-century AD Marcus Aurelius (1995), fifth-century AD Boethius (2004), and Descartes ([1641] 1988) write about illness and its contribution to the modes and themes of philosophizing. These thinkers also reflect on the relationship between health and virtue and health's contribution to the good life, as I discuss below.[5] We find Descartes commenting in his 1637 text, *Discourse on Method*:

> For even the mind depends so much on the temperament and disposition of the bodily organs that if it is possible to find some means of making men in general wiser and more skilful than they have been up till now, I believe we must look for it in medicine (1988, p. 47).

Descartes sees health as 'the chief good and the foundation of all other goods in this life' (p. 47). Perhaps most famously, Montaigne ([1580] 1993), following Socrates, claims that the whole point of philosophy is to prepare us for illness and ultimately death.

Philosophical reflection on illness in the Western tradition has tended to be shaped by Stoic, Epicurean, and, later, Christian philosophies, each of which emphasize the importance of achieving reflective coping with illness, seen as an essential feature of the world. Why this reflective attitude to illness is essential differs by tradition. The Stoics seem to

---

[4] See, for example, the Oxford University Press series Biographies of Disease <http://ukcatalogue.oup.com/category/academic/series/medicine/bod.do> (accessed 21 October 2014).

[5] See Boethius' *Consolation of Philosophy*, Book IV, Seneca's *On the Shortness of Life*, Epictetus' *Discourses*, and Marcus Aurelius' *Meditations*.

argue that everything that exists, including ostensibly bad things such as illness, are essential components of the rational order of the cosmos, so the properly philosophical response is to recognize this and reflectively accept illness. We find Epictetus (2004) saying: 'A man who has a fever may say: If I philosophize any longer, may I be hanged: wherever I go, I must take care of the poor body, that a fever may not come. But what is philosophizing? Is it not a preparation against events which may happen?' (*Discourses*, 'In what manner we ought to bear sickness').

But for later Christian thinkers, such as Boethius, illness is a mark of our corrupt, imperfect state, and hence not an original feature of God's design. Boethius characterizes wickedness of the soul as akin to bodily sickness; while the wicked deserve hatred, the ill should be treated with pity (*Consolation of Philosophy*, Book IV). So the properly philosophical response is to use illness in a doubly edifying way: first, as a reminder of the frailty and corruption of our mortal status and, second, as a source of moral and spiritual improvement (Kidd 2012).

This is now largely a lost theme in philosophy because of the gradual erosion of philosophy's phronetic role (although see Nussbaum 1994).[6] I propose that more work needs to be done to examine and describe the philosophical role of illness. In this chapter I outline some of the ways in which illness is philosophically relevant (see also Kidd 2012). I suggest that illness is relevant to philosophy because it uncovers aspects of embodied existence and experience in ways that reveal important dimensions of human life. It does this by broadening the spectrum of embodied experience into the pathological domain, in the process shedding light on normal experience, revealing its ordinary and therefore overlooked structure. Illness broadens the range of bodily and mental experience (e.g. delusions, dementia). Moreover, at present illness is an integral part of biological life and thus must be taken into account when considering human life as a whole.

Discussions of the good life, human relationships, and ethics would be incomplete if they did not take into account the full spectrum of human

---

[6] There is much discussion in the philosophy of medicine about the concept of illness (and disease), and its relationship with the concept of health. But this conceptual analysis does not touch on the existential or philosophical role illness may have. For a notable exception see Toombs (1993).

life and experience, spanning sickness and health, childhood, adulthood, and old age, and the arising complexity of notions such as autonomy and choice. In addition, illness is an opportunity for reflection, because of its distancing effect, which illuminates taken-for-granted values and expectations by destroying the assumptions that underpin them (e.g. assumptions about longevity, the good life, and autonomy). I suggest that these characteristics warrant illness a philosophical role.

However, illness is a unique form of philosophizing. While the execution of most philosophical procedures such as casting doubt or questioning is volitional and theoretical, illness is uninvited and threatening. Illness throws the ill person into a state of anxiety and uncertainty. As such it can be viewed as a radical, violent philosophical motivation that can profoundly alter our outlook. I argue that the radical nature of illness should be utilized in order to both sharpen and expand philosophical discussion.

In addition, illness is a form of invitation to philosophical reflection that arguably has a more universal and intimate character than other forms of invitation. For example, the Platonic 'sense of wonder' as the basis for philosophical impulse is something that many people do not experience. Similarly, metaphysical and existential curiosity is a trigger to philosophize that is not shared by all. But everyone will die and most will get sick and age. So at the very least, these processes are invitations to reflect on these facts about one's own life. Not everyone cares about truth or beauty or the nature of reality, but almost everyone cares about living a good life, even if that means a life of coping—reflectively and practically—with the loss, pain, and suffering that illness brings.

I conclude the chapter by examining how illness may impact upon the practice of philosophy. I argue that illness can be integral to philosophical method in a number of ways: in shaping and influencing philosophical methods and concerns, modifying one's sense of philosophical salience and conception of philosophy, and increasing the urgency and appeal of particular philosophical topics. Section 9.1 outlines the centrality of the body for human experience and discusses how illness changes embodiment, meaning, and being in the world. Section 9.2 discusses illness as a form of suspension, or *epoché*, performed through objectification and uncanniness. Section 9.3 discusses illness as a motivation to philosophize and outlines how illness may change our modes and styles of philosophizing.

9.1 Illness Modifies Embodiment, Meaning, and Being in the World

Let us start by recapping some of the ideas discussed earlier in the book. Three aspects of existence are significantly modified by illness: embodiment, meaning, and being in the world. Embodiment is the fundamental feature of human existence (Merleau-Ponty 2012; Clark 1997, 2008; Wheeler 2005). Cognition and behaviour cannot be accounted for without considering the perceptual and motor apparatus that facilitates our dealing with the world (Calvo and Gomila 2008, p. 7). The body is the condition of possibility for perception and interaction with spatial objects and our means for having a world. As Gallagher and Zahavi (2008) write 'the body is considered a constitutive or transcendental principle, precisely because it is involved in the very possibility of experience' (2008, p. 135). Every worldly experience is mediated and made possible by embodiment (Zahavi 2003, p. 99). Or as Merleau-Ponty (2012) put it, the body is 'that which causes [things] to begin to exist as things under our hands and eyes' (p. 146).

Counter to a purely naturalistic understanding, the body is not merely a thing among things. Embodiment determines spatial relations and temporal experiences, while also participating in these relations as a secondary form. The body is 'the centre around which and in relation to which space unfolds itself' (Zahavi 2003, p. 99). According to Husserl, motility and tactile experience are fundamental not just for perception but for any organized subjective experience (Husserl 1997). In this sense the body is the foundation of human experience. As Taylor Carman writes, the body 'plays a constitutive role in experience precisely by grounding, making possible, and yet remaining peripheral in the horizons of our conceptual awareness' (1999, p. 208). Or to use Merleau-Ponty's famous formulation, the body is 'our general medium for having a world' (1962, p. 146).

The form of my embodiment serves as part of the background of my experience (Smith 2007, p. 223). This structure defines, for example, the coordinate system of my visual field and my proprioception. Different sensory fields are bound together to create a unified stream of meaningful experiences, united by a body with an established repertoire of habits, activities, and style (on style, see Meacham 2013). In Husserl's terms, the constitution of my body is essential to the constitution of objects appearing to me and indeed to the constitution of space and time (Husserl 1997, §73).

Given how central the body is, a change to a bodily function entails a change to one's way of being in the world. Such a change will also affect the meaning of experience. For example, the experience of dancing will be radically altered by respiratory disease, both on the level of bodily feeling, which turns from a pleasurable experience to one of exertion, and on the level of meaning, when it changes from an experience of 'I can' to an 'I cannot' (compare Carel 2012). The types of changes affected by illness may vary greatly, from changes to sensory experience and meaning to cognitive and emotional experience.

If we think about symptoms as disparate as loss of mobility, loss of memory, and incontinence, we can see that such changes are radical and remove the ill person from the realm of familiar, predictable, and well-understood experience. This displacement from the familiar destabilizes the structure of experience and reveals new aspects of our being, such as our ability to adapt, mourning, and dependency. The bodily foundations of autonomous adulthood are often removed, revealing the tentative and temporary nature of these foundations. Illness plays an important existential role: it can disclose finitude, dis-ability, and alienation from one's body as extreme modes of being.

The philosophical illumination offered by the study of illness is explored by Matthew Ratcliffe, who studied the experience of time in depression (2012b). Ratcliffe argues that there is strong evidence that the experience of time is affected in a number of ways in depression. He offers a phenomenological analysis of this experience, using Thomas Fuchs's application of Husserl's notion of retention and protention to the experience of time in depression. On this account time both slows down and accelerates in depression. This alteration to the normal experience of time can be explained by the effects of depression. On Ratcliffe's account, depression removes meaning, obliterates the desire to carry out projects, and stops the attribution of value to different projects in the depressed person's world (2012b). Ratcliffe claims that the breakdown in such cases is not merely in the *contents* of experience but in the *structure* of experience itself.

Because illness can affect many body parts and functions, it can delineate different aspects of embodiment by serving as a limit case

(Carel 2013b).[7] The loss in illness may be of overall functionality, but also of flexibility and variability. With a narrowed spectrum of activity, one's motility, assessment of effort and duration, and notions such as 'difficult' and 'far', are modified. The restriction is not only a conscious understanding but underlies the kind of action one's body spontaneously performs. Here is a description of such pre-reflective modification:

Every time I tried—and failed—to do something that was too strenuous my body stoically registered the failure and thereafter avoided that action. The change was subtle, because this happened by stealth [...] I stopped feeling all the things I could not do. They were quietly removed from my bodily repertoire in a way so subtle I hardly noticed it (Carel 2013a, pp. 40–1).

Illness may lead to a collapse of meaning, or what Heidegger calls anxiety (1962). In anxiety one's overall sense of purposeful activity is lost, leaving the person experiencing anxiety unable to act. Action is grounded in meaning: I pull a shirt over my head in order to get dressed. I get dressed in order to go to work. I go to work in order to earn a living, and so on. Ultimately, this nested set of goal-directed activities comes to an end and human existence is ungrounded. A realization of the groundlessness of human existence leads to what Heidegger calls anxiety (*Angst*).

In anxiety purposefulness disappears and the meaning of entities is lost. They turn from being ready-to-hand (*Zuhanden*) entities we use (t-shirt, shoes, reading lamp) to being present-at-hand (*Vorhanden*) entities which confront us with their lack of usefulness, and hence their lack of meaning. In anxiety intelligibility is lost because the practical coherence of entities has been lost with the sense of purposefulness.

Illness can also give rise to another kind of loss of meaning, related to the loss of the ability to perceive things as useful tools, and experiencing the contingency and irretrievability of meaning. In somatic illness a ready-to-hand entity like a staircase can turn from being a practical tool to a present-at-hand entity, or even a conspicuous obstacle. Toombs, a philosopher suffering from multiple sclerosis, writes: 'the bookcase outside my bedroom was once intended by my body as a "repository for books"; then as "that which is to be grasped for support on the way to the bathroom", and is now intended as "an obstacle to get around with my wheelchair"' (1995, p. 16).

---

[7] Death would not be the ultimate limit case, but crossing the limit.

Somatic illness may cause a sudden and often disturbing sense of the contingency of the meanings and uses we assign to things: 'The bookcase holds books. Of course it does! What else might it do? It might obstruct, impede, sadly remind . . . '. There is also a sense of the irretrievability of certain meanings: 'the bookcase will always be an obstacle and will only cease to be so once I cease to be so'. The sense of inhabiting a space of possibilities can be replaced by a sense of this space becoming delimited and static. The changes brought about by illness are not localized to a specific object, but modify one's entire interaction with objects and the environment, i.e. their being in the world. For a wheelchair-user it is not just this shop or that doorway that are inaccessible, but the environment as a whole becomes less inviting or even hostile. Illness can expose not only the limits of human existence but also the biases of an environment.

Illness may be philosophically salient in one of two ways. It is, in some cases, a severe and sudden disruption of our life. In this situation the illness is something foreign, threatening, and disruptive which we seek to get rid of. A bout of flu or gastric infection are examples of this type of illness. This type of illness is philosophically useful because of its acute disruption of the everyday; it makes visible the taken-for-granted manner in which we structure our routine life. We take for granted that we can plan our day, do things, and get from one place to another. These tacit assumptions are placed in abeyance in the case of a sudden illness. Feelings of missing out, being useless, and feeling unwell expose the underlying sense of participation, purposefulness, and potency that has been disturbed.[8]

Illness may also appear more subtly. Symptoms may be minor and not quite noticeable until they reach a certain threshold or are picked up in routine screening. In this case the illness is not an acute disruption of the everyday, but still alters the everyday capacities of the ill person, and thus may also give rise to philosophical reflection, albeit of a different sort. Shaun Gallagher describes this kind of illness as one that 'either sneaks

---

[8] The experience of ageing may also give rise to these sensations, but more gradually than sudden illness. One's view on this will depend upon the conception of ageing one operates with. A gerontophobe will find that ageing dims their potency in a way that distresses them, but a gerontophile, such as Cicero or Montaigne, will welcome this by interpreting the loss of potency as Nature's way of teaching one to withdraw from the vigorous pursuits of youth and instead devote oneself to study, reflection, and gardening. I thank Ian James Kidd for making this point.

up on us, or that we become so habituated to (perhaps because it won't go away) that it defines our form of life—it becomes us, or we become it' (unpublished presentation 2009). Whereas in acute illness the expectation that the illness will 'go away' is very much part of the experience of illness, this expectation disappears in chronic illness. Arthur Frank (1991) contrasts his heart attack, which he interpreted as 'an incident', with his cancer:

After an incident like my heart attack I was able to bounce back [. . .] That's accurate because in most cases we do not sink into an experience, we only hit the surface. I may have bounced back from a heart attack, but with cancer I was going to have to sink all the way through and discover a life on the other side (1991, p. 28).

The second type of illness is not a disruption, but a 'complete form of existence', as Gallagher writes, following Merleau-Ponty (Gallagher 2009; compare Meleau-Ponty 2012). In this case, the disturbance runs deeper and longer, and thus must be dealt with in a different way than a passing illness such as food poisoning. When illness becomes a complete form of life, concepts such as 'worthwhile' or 'difficult' are modified, the expectations the ill person has of her life change, and her understanding of time and value is readjusted in light of her prognosis. Chronic or progressive illness is a comprehensive realignment of meaning, values, and ways of being that culminates in illness becoming one's complete form of existence. This process is a kind of distancing from one's previous form of existence, and as such it opens it to philosophical examination.

9.2 Illness as *Epoché:* Objectification and Uncanniness

Because illness removes the taken-for-granted nature of motility and bodily capability, it makes what is normally natural and unreflective artificial and conscious (Gallagher 2005). In this section I explain how this shift from tacit and effortless to explicit and effortful embodiment drives philosophical reflection. In this shift from health to illness bodily experience not only becomes explicit, but also becomes negative and characterized by objectification and uncanniness. These two notions are used here to demonstrate the role of illness as a mode of philosophizing.

Illness can be seen as a crisis of meaning in one's life. This crisis arises from a collapse of the ill person's life narrative (Williams 2003) but also a disruption of routines, habits, expectations, and abilities. This disruption shakes one's everyday life and provides a distance from it. This distance has been described by Arthur Frank (1991) as a 'dangerous opportunity':

Critical illness offers the experience of being taken to the threshold of life, from which you can see where your life could end. From that vantage point you are both forced and allowed to think in new ways about the value of your life. Alive, but detached from everyday living, you can finally stop to consider why you live as you have [. . .] (1991, p. 1).

This brings to mind the ancient Greek conception of philosophy—introduced by Socrates and embraced by the Stoics, and later valorized by Montaigne—that to philosophize is to learn how to die (Montaigne 1993). In this context, learning how to die means more than accepting one's mortality. It furnishes this highly abstract demand with concrete content. Learning how to die means learning to bear illness well, gaining the ability to confront pain and disability, accepting one's diminishing abilities, and dealing with difficult emotions such as mourning, envy, and sadness. This demanding task may underpin the process of post-traumatic growth, and explains why encountering diverse forms of embodiment and life can lead to personal growth, moral development, and increased sensitivity to suffering. It also calls for the correct response, which I call reflective coping. Such coping may enable the ill person to bear their illness well. An exemplary outline of such reflective coping can be found in Epictetus: 'What is it to bear a fever well? Not to blame God or man; not to be afflicted at that which happens, to expect death well and nobly, to do what must be done' (*Discourses*, 'In what manner we ought to bear sickness').

As discussed in earlier chapters, illness calls upon the ill person to explore her life, its meaning, priorities, and values. This personal quest is well documented in sociology of medicine, medical anthropology, qualitative healthcare research, and cancer psychology (Brennan 2012; Thorne and Paterson 1998; Thorne et al. 2002). But illness is also a distinctively philosophical tool motivating reflection, by moving beyond the idiosyncratic and personal to a general and abstract exploration of embodiment as a source of meaning and condition of possibility for the self.

In particular, the anxiety, loss of meaning, and defamiliarization described in the previous section give rise to a peculiar form of what

Husserl termed the *epoché*, a suspension of our 'natural attitude'. The *epoché* asks us to dislodge ourselves from everyday habits and routines in order to reflect on them. This, I suggest, is what happens in illness, albeit in a raw and unformulated manner.[9] Illness is a particular form of *philosophical motivation* forced upon the ill person and characterized by violence and negativity.

The *epoché* asks us to shift our focus from objects to acts of perception, but does not involve ceasing to perceive; it is not a sceptical procedure. It is a shift in one's way of being in the world that enables philosophical reflection without ceasing to be part of the world. Exercising the *epoché* involves stripping away shared meanings and familiar connections between person and object. The object then becomes freed from tacit and accepted modes of perceiving and understanding it, and appears in novel ways. Thus the experience of illness, or anxiety, as a particular type of *epoché* can shed new light on taken-for-granted aspects of the world.[10]

Illness suspends the natural attitude: the accepted meaning-laden, metaphysically determined way of experiencing the world. Such suspension does not mean doing away with the natural attitude, which is impossible, but maintaining the attitude while suspending the underlying beliefs underpinning it. This is the neutralization of one's belief in the existence of the world or of an object, which Husserl called the *epoché*. This neutralization is employed in the shift from the natural to the critical attitude (Drummond 2007, pp. 67–8).

We do not affect the *epoché* in order to 'deny, doubt, neglect, abandon, or exclude reality from our research, but simply to suspend or neutralize a certain dogmatic attitude toward reality [...]' (Zahavi 2003, p. 45). Bracketing the natural attitude is a withdrawal from the ordinarily implicit commitment to the reality of the world (Ratcliffe 2008a, p. 4). Bracketing turns the world into a phenomenon of being, instead of something that is.

---

[9] Distancing can also arise as a result of other life events such as bereavement, divorce, and other trauma.

[10] It is likely that other life events, such as grief, accident, moving to a new country, or the loss of a long-term relationship facilitate the same kind of phenomenological bracketing, by eroding something that was previously taken for granted, and thus drawing one's attention to that aspect of experience. I thank an anonymous referee for making this comment.

As Husserl makes clear, this is not a sceptical or idealist position. Rather, this 'inhibiting' or 'putting out of play' of the natural attitude exposes 'my pure living [...] the universe of phenomena in the phenomenological sense' (Husserl 1988, p. 20). This suspension neither questions nor negates reality; rather, it allows under-theorized aspects of experience to become an object of enquiry, because it enables us to shift attention from the given object to the way in which it is given and its modes of appearance. As Husserl writes in *Ideas I*, 'the whole prediscovered world posited in the natural attitude [...] is now without validity for us; without being tested and without being contested, it shall be parenthesised' (1982, p. 62). But importantly, the *epoché* 'leaves everything exactly as it is' (Smith 2003, p. 23).

Zahavi characterizes the *epoché* as a philosophical 'entry gate' (2003, p. 46). I suggest that because of its defamiliarizing and distancing effect illness is such an entry gate into philosophy. It is an invitation to investigate subjectivity in illness and thus to expand the conditions under which subjectivity is studied. It can reveal novel facets of subjectivity that otherwise remain unnoticed. For example, let us return to the case of Schneider discussed in earlier chapters (Merleau-Ponty 2012, p. 103ff.). Merleau-Ponty interprets Schneider's inability to perform abstract movements, initiate sexual relations, or stray from a daily routine as the breakdown of his intentional arc. He uses this case study to illuminate how this intentional arc operates normally, something that would go unnoticed. Similarly, Shaun Gallagher (2005) discusses the case of Ian Waterman, who suffered from de-efferentation from the neck down. Waterman was forced to use vision to locate his limbs and identify his posture, because of the loss of sensation and proprioception. Gallagher uses Waterman's case to provide an account of normal proprioception. The breakdown of normal human existence provides a unique opportunity to uncover facets of existence that are not normally visible.

In illness, the *epoché* is *forced upon* the ill person, because of the modification to and limitation on her body imposed by illness. The ill person may have no interest in philosophy and no desire to undergo existential change. However, illness—an uninvited guest—forces itself upon the ill person, and compels her to modify and thus re-examine her bodily habits, existential expectations, experience of body, space and time, and way of being in the world (Carel 2012). Illness is a form of violent suspension of the natural attitude, which enacts a philosophical procedure in a way that is far more brutal than normal philosophical reflection.

Illness motivates ill people, and often those around them, to confront practical concerns, and this, in turn, gives rise to theoretical reflection on one's embodied situation. It is an uninvited type of reflection, but such coping with practical concerns reveals the normal conditions under which one previously operated in health. It replaces health, which is 'life lived in the silence of the organs', as the French surgeon Leriche wrote (cited in Canguilhem 1991, p. 91). Illness allows these conditions to be explored, as their seamless function is lost and they become the object of explicit attention. The natural attitude is not immune to theorizing or meta-reflection under circumstances which disrupt it. Illness is one such circumstance.

Merleau-Ponty characterizes the *epoché* as an experience of 'wonder in the face of the world' (1962, p. xiii). This sense of wonder, interrogation, puzzlement, characterizes some experiences of illness. For example, it drove Randy Pausch (2008) to write *The Last Lecture*, a series of talks about life and death, after being diagnosed with pancreatic cancer. 'Many people might expect the talk to be about dying. But it had to be about *living*', he writes (2008, p. 9). Because of changes to the somatic or mental architecture of one's body (or mind), one's contact with and experience of the world is radically modified in illness. One's sense of comfort and familiarity may be displaced by alienation and a sense of 'not being at home' (Svenaeus 2000a, 2000b).

Merleau-Ponty writes: '[Reflection] loosens the intentional threads that connect us to the world in order to make them appear; it alone is conscious of the world because it reveals that world as strange and paradoxical' (2012, p. xxvii). I suggest that illness is such a slackening of the intentional threads, which reveals the world and embodiment as uncanny. Illness problematizes the relationship to one's world, thus inviting or forcing philosophical reflection.

The *epoché* also arises from the rift between the biological and lived body, which becomes observable in illness. In health the two aspects of the body usually cohere, or respond harmoniously to a range of experiences (but see Chapter 2). In illness the biological body comes to the fore, as it ceases to cooperate with the ill person's desires. For example, a diabetic's biological body will be unable to cope with a chocolate mousse, despite her lived body's craving for it. In addition to the rift, the biological body also becomes the source of pain, disability, and failure. In this respect it becomes the source of negative experiences and the focus of medical attention, which often further distance us from it (Carel 2013a).

Lawrence Hass views illness as conflict between the biological body and life projects. While the individual person's 'personal life' is engaged in a project, the biological body obstructs it. For example, one's personal aim may be to become a parent. However, if the biological body is infertile, the result is a clash between the desire to have a child and the biological barrier. The 'impersonal operations' of the biological body, over which we have little or no control, interfere with the intentional arc of the person, the meaningful connection between person and world that is aimed at a particular goal (Hass 2008, p. 87). The sense that one's body is an obstacle, a problem, something that is no longer well-understood, may initiate a kind of *epoché*. The metaphysical status of the body is thrown into question, because it is no longer familiar and predictable. In other words, the body is subject to a process of objectification in illness, as well as becoming uncanny—two processes to which we now turn.[11]

Objectification— The natural process secondary to experiencing the lived body is experiencing the body as an object amongst objects. In illness this process takes on a new dimension, as modern medicine and the sciences underpinning it view the body as a physical object.[12] This objectification takes place under the dual experience we have of our bodies. The body is experienced as both a lived, pre-reflective body (my first-person experience of and through it) and as an objectified, observed, spatial object (the third-person experience of it) (Merleau-Ponty 2012; Sartre 2003). It is both a physical object and the seat of consciousness.

---

[11] Young, healthy embodiment is typically oblivious to the possibility that the body might be experienced in this way. Confidence in one's physical and cognitive capacities can occlude a sense that these capacities might change (even with natural ageing) and that this change will increasingly come to radically impact one's identity. This can be seen as a failure of moral imagination, compassion, humility, or even a misunderstanding or denial of the biological expiration that delimits human life. I thank Ian James Kidd for raising this point.

[12] Pragmatically this is a good thing. Modern medicine has made huge progress because of this view.

The exploration of objects implies a simultaneous self-exploration and self-constitution; there is a reciprocal codependency between the processes. 'The world is given to us as bodily investigated, and the body is revealed to us in this exploration of the world' (Zahavi, 2003, p. 105). We are aware of perceptual objects because we are aware of our bodies and how the two interact. When we investigate objects, this is always accompanied by some kind of bodily self-awareness. In illness objectification gives rise to a distance between oneself and one's body, which is now reified into an object of medical inquiry and treatment. Objectification breaks down the natural taken-for-granted attitude towards the body, the seamless unity between the body as object and the body as subject.

Merleau-Ponty claims that the body is the first object we perceive *as* an object, thematizing and learning to interpret and judge it according to cultural standards (Merleau-Ponty 2012). Prior to that event, I do not experience my body; rather, I experience *through* my body. As Zahavi writes, 'Originally my body is experienced as a unified field of activity and affectivity, as a volitional structure, a potentiality of mobility, as an "I do" and "I can"' (2003, p. 101). Illness impedes the natural sense of ability and activity, revealing the volitional structures of embodiment. Our natural orientation is one in which the body serves as the perceptual centre of our experience, with our attention directed away from it, rather than to it. The negative, unwanted focus on the body in illness reorients our attention back towards the body, but this time viewed as an object. Many of us have had the experience of seeing an X-ray or scan of our bodies and having to relate our subjective feeling of our body to this objectifying image.

The duality of the body plays a complex role in healthcare provision. The health professional experiences the patient's body as an object, but is also aware of its subjectivity (so will apologize for having cold hands when touching a patient). The patient may feel objectified by the physician's gaze, but this objectification is only possible because she is first a subject (Carel and Macnaughton 2012). The physician perceives an appearance of an experienced object: a swollen arm. The patient perceives a localized sensing: the sore arm. She may also be shown an X-ray of her arm, and will thus oscillate between the two experiences—the immediate pain localized in the arm, and the arm as an object that is gazed at and imaged. She can focus on the sensing (observing the swollen arm) or the sensed (the arm itself), and each will yield a differently thematized experience.[13] Health professionals often view the body as thematized and objectified, focusing on a particular organ or function in order to understand it as a medical object. But for the patient, the awareness of her body as an object is secondary to her subjective experience of receiving healthcare.

As Fredrik Svenaeus claims, modern medicine expands the objecthood of the body through imaging and conceptualization of organs, functions, and molecular processes (2012). The medical emphasis on the objecthood of the body contributes to the rift between the body as lived and the biological body. This intense experience of the objecthood of the body in illness alienates the patient from her body. Jean-Dominique Bauby (2007), who suffered a stroke that resulted in locked-in syndrome, writes:

Reflected in the glass I saw the head of a man who seemed to have emerged from a vat of formaldehyde. His mouth was twisted, his nose damaged, his hair tousled, his gaze full of fear. One eye was sewn shut, the other goggled like that doomed eye of Cain. For a moment I stared at that dilated pupil before I realized it was only mine (2007, pp. 32–3).

As this passage shows, illness may force us to adopt a reifying and detached view of our own body—this is a shift often required from patients when discussing their disease with health professionals. However, although most of us can momentarily adopt an objective view of our body, we are not able to sustain it; that is existentially unbearable. We cannot actually view ourselves *objectively* in any sustained sense, and it is unrealistic to expect that of others. Health professionals need to be aware of this because of medicine's privileging of third-person perspectives. Objectivity is seen as an ideal by many health professionals, but when subjected to philosophical analysis, it can be seen that merely relying on an objective stance is a naïve and non-practicable position that ought to be replaced with a more nuanced understanding of intersubjectivity.

A further objectification takes place in the clinic. When a patient awaits her blood test results, she is as ignorant about her cholesterol levels, for example, as an objective observer. When she asks the physician 'how bad is it?' that is because she is genuinely unable to access this information by examining her bodily sensations. The patient's body is an

---

[13] The health professional may also alternate between the sensing (her experience of gazing at the X-ray or examining the arm) and the sensed (the arm or the X-ray), but this oscillation does not involve self-objectification.

object, not only to the physician, but also to the patient herself. Other experiences of objectification can be seen in encounters with medical technology. Seeing one's tumour as a set of CT images or aligning your limbs for a bone density scan can make the objecthood of the body prominent in one's experience. These objectifying experiences may lead to a sense of alienation from one's body, and to treating that body as an aberrant object over which one has little control. The ill body becomes despised, feared, and alien.

However, this objectification is not complete. There is an oscillation between treating one's own body as an object of medicine and the subjective experience of apprehension, feeling cold, or flinching from the physician's touch. Husserl's example of two hands touching each other makes this duality salient (Husserl 1988). When the right hand is the active, touching one, it is at the same time being touched by the left hand. If we consciously decide to reverse the roles and concentrate on the left hand as touching, we still oscillate between both dimensions, the active touching one and the passive dimension of being touched.

As discussed in Chapter 2, section 2, this duality of experience is a unique feature of human existence. In order to touch, one has to be a thing among things, a physical object. As such an object, one has to be open to the possibility that one can be touched. However, in illness the natural movement between the two dimensions is disrupted because the passive dimension becomes prominent. For example, internal examination gives rise to an experience of being touched from within, thus expanding the domain of passivity. The body as object takes precedence in the clinical context, and its foreignness is accentuated by the inaccessibility of some medical facts to the patient other than via a third-person report. In illness one's body becomes an object in ways it would not otherwise have.

Uncanniness— In illness the body becomes an obstacle and a threat, instead of my home, a familiar place I inhabit. A change to one's body is a change to one's sense of being at home in the world. The body ceases to be the 'null centre' of my orientation towards the world (Smith 2003, p. 221) and instead becomes the source of negative experiences. The primitive sense of 'I can' is replaced by a conscious, artificial, mediated sense of 'I cannot' (Kesserling 1990). The perspicuous nature of bodily orientation as being the foundation of experience becomes occluded with attention.

Illness can suspend the familiar setting and feelings that underpin normal everyday actions, giving rise instead to an experience of 'unhomelike being in the world' (Svenaeus 2000a, p. 9; 2012). Uncanniness arises most forcefully from the disruption of this background, which happens as a result of changed embodiment. Our concepts, habits, routines, expectations, and norms may be disrupted or even destroyed by illness. Uncanniness arises from a new, negative focus on one's body, a sense of this body becoming an alien destructive force, or even the threat of annihilation. This changes the ill person's relationship to her environment, as well as her concepts.

Illness causes disruption of the lived body, which interrupts the relationship between one's body and the environment. As discussed in earlier chapters, concepts such as 'near' and 'easy' change their meaning for the individual, who may experience a further sense of alienation because her new use of concepts moves away from the norm. In addition, such concepts acquire new *objects*, e.g. routine activities such as carrying a laptop bag, or nipping upstairs become marked as difficult in illness. Many concepts change their meaning, as well as attaching to new objects, and so changing in scope.

The change is not merely linguistic; the ill person actually *experiences* the physical world as less welcoming, full of obstacles, difficult. Distances increase, everyday routines take up more time, activities have to be forsaken or redesigned, and so on. Toombs describes loss of mobility as 'anchoring one in the Here, engendering a heightened sense of distance between oneself and surrounding things' (1990, p. 11). Illness modifies not only one's body, but one's sense of space, as was discussed in Chapter 3.

Not only the experience of space and the use of concepts change in illness; the experience of time may also change and contribute to the sense of alienation and uncanniness brought about by bodily changes, fear, pain, and limitation. Sustained pain or a poor prognosis may completely transform one's experience of time (Toombs 1990). Activities may take more time, and thus expand, or may become impossible, which may cause the ill person to experience herself as 'useless' or as more disabled than she is (Toombs 1988). Insecurity and anxiety about future health and ability may make one focus on the present (Carel 2013a, ch. 5).

And memories of a healthy past become objects of regret, yearning, or a sense of discontinuity (Bury 1982).

The experience of time may also radically change in response to an uncertain prognosis. Priorities might change and it is an opportunity to question how one has lived and how one would like to live (Lindsey 1996; Lindqvist et al. 2006). These changes are fundamental and may lead to shifts in identity (Williams 2003) as well as triggering philosophical questions. How plastic is the experience of space and time? What determines 'normal' experience? Can there be continuity in identity and personhood given the radical change in one's experience? The way in which such questions arise as a consequence of bodily change in illness demonstrates that illness can motivate, or serve as an entry gate to, philosophical activity. We now turn to examine the philosophical salience of such motivating.

9.3 Illness as Invitation to Philosophize

So far I explained how illness can be philosophically illuminating, by disrupting everyday taken-for-granted assumptions about embodied existence, and thus performing a kind of *epoché*.[14] In this section I look more closely at this process, and suggest that illness is a peculiar kind of motivation to philosophize.

Illness is unwanted; it is almost never welcome or easily accepted into one's life. It is also a radical event: it gives rise to a rethinking of values and meaning, given the changed life conditions. As articulated throughout the book, serious illness is a dramatic life event that affects all aspects of life. Because of these features illness can *motivate* philosophical reflection. However, it is not only that illness can motivate the person who falls ill to become more reflective, although as can be seen through this book this is certainly true.[15] More importantly, the features that motivate reflection in individuals who become ill make illness salient to the practice of philosophy.

---

[14] This process may affect family members or carers who become distanced from shared practices and understandings by the limitations of illness although they do not experience it first-hand themselves.

[15] This is not necessarily the case: illness can also be met with denial, repression, or other defence mechanisms that eschew reflection.

Illness invites or inspires reflection of a philosophical sort. But it can also brutally force this reflection on ill people; for example, consider the way a poor prognosis forces the ill person to confront their death. It forces the ill person to consider death not in the abstract—a luxury of the healthy and young—but in its most intrusive and frightening. Illness does not permit an inauthentic attitude towards death as an abstract, far-off event. Illness forces the ill person to face her own death in a most concrete way. Illness is a strict philosophical instructor forcing the ill person to confront death in its most immediate. This can be seen as a fuller, more existentially salient form of philosophizing. Indeed, for Heidegger, authentically facing death demands precisely this kind of first-person engagement with one's mortality.

Illness is also different to other motivations to philosophize. Whereas normally one chooses to perform a philosophical procedure, of say, questioning or criticizing an argument, illness is violent, unwanted, destructive, and uncontrolled. We normally take the practice of philosophy to be a matter of choice, whereas illness is rarely chosen. We think of reflection as a pleasant experience of intellectual challenge; but the reflection prompted by illness is all-consuming, extreme, and terrifying.

Illness affects different aspects of philosophical reflection: it can call for more radical and personal methods, such as existentialism or nihilism. It affects the philosophical concerns of the ill person—issues such as death, the good life, causation, and time can be central and pressing for ill people in a way they would not otherwise be. Because it forces the ill person to engage with their physical or mental decline and death, it triggers reflection on finitude, dis-ability, suffering, and injustice. Similarly, the urgency and salience of particular philosophical topics may change in light of illness (e.g. Seneca's On the Shortness of Life (2004)). The very activity of philosophizing may change and become more urgent and personal.

Illness may also change the ill person's conception of philosophy (if she has one) as a vital practice aimed at a good life, rather than a theoretical enquiry, for example. The former view of philosophy can be seen in ancient philosophical schools such as the Epicureans and the Stoics, and is perhaps in part a response to the period's limited ability to treat illness and control pain. With such limited ability comes a need to modify one's attitude to illness; this is dramatically different from the 'treat-at-all-costs' attitude fostered in many quarters of modern medicine (Gawande 2015).

Illness may also bring about the sense that philosophical enquiry ought to be integrated into, and so is intrinsic to, one's life as a whole. A case in point is Alasdair MacIntyre (1999), who stresses the fact of our vulnerable, dependent, afflicted state as a precondition for a style of moral philosophizing attentive to the human condition. Illness is particularly important to ethics, for these MacIntyrean reasons: in order to understand virtue, goodness, and compassion we have to look at those experiences in life that really call them into play, even test them, such as being ill, caring for the ill, or empathically engaging with ill people. However, much work in ethics still ignores the fact that people get sick and that responding to them well requires a range of virtues such as compassion, empathy, patience, and courage.

And finally, as Ian James Kidd suggests (personal communication), even metaphysics can look to illness for instruction. Illness can teach us (or even enact in our own bodies) certain truths about the nature of reality. Buddhists, for example, could interpret experiences of somatic illness as instructive not only of certain moral truths, but certain metaphysical truths—e.g. the conditioned nature of all phenomena, and the subjection of all things to impermanence and flux (*anicca*).

However, illness does not always or necessarily fulfil its role as inviting to philosophize. It is disorientating and overwhelming and can, like other extreme hardships, destroy reflection instead of bringing it about. Illness is not philosophical reflection in itself, but can be—and often is—a *way into* reflection. It is a compulsive invitation to philosophize. It is compulsive because it is unchosen, but it is an invitation because it is not always taken up. This compulsive invitation has been described in many pathographies (Frank 1991; Pausch 2008):

The experience of illness and its sweeping effect on every aspect of life shocked me into thinking about these issues. I found that I had to reinvent my life [. . .] I learned to rethink my aspirations and plans. I relinquished the sense of control I previously had [. . .] My experiences pushed me to reflect on health and illness (Carel 2013a, p. 9).

9.4 Conclusion

True philosophy, Merleau-Ponty says, 'entails learning to see the world anew' (2012, p. xxxv). Illness forces us to learn not just to re-examine the world, but also to cope with it, negotiate new limitations, and continue to live to the best of our ability within the constraints of illness. The outcome of such coping with practical limitations can be existential and philosophical illumination. Perhaps illness is a kind of philosophical method, which illuminates normalcy through its pathological counterpart. However, Merleau-Ponty calls on us to make this claim carefully:

It is impossible to deduce the normal from the pathological, deficiencies from the substitute functions, by a mere change of the sign. We must take substitutions as substitutions, as allusions to some fundamental function that they are striving to make good, and the direct image of which they fail to furnish (2012, p. 110).

Merleau-Ponty is aware that the pathological is not merely 'a change of the sign'. Rather, pathological cases allude to some function they are 'striving to make good' and in this striving create a complete form of life. It is this completeness that invites further philosophical investigation to reveal how what may seem pathological and deficient may give rise to phenomena such as adaptability (Carel 2007) and edification (Kidd 2012). Canguilhem (1993) defined disease as 'a new way of life for the organism', the creation of new norms that govern the relationship of the diseased organism to its environment (p. 84). The richness of the experience of illness and the understanding of health and illness as distinctly *normative* attest to the fact that illness both requires and merits further philosophical exploration.

Among the philosophers of the ancient Greek, Indian, and Chinese traditions, philosophy was taken to be a set of vital practices aimed at the good life, part of what Pierre Hadot called 'a way of life'. Philosophy was a disciplined system of ethical transformation that proceeded, typically, from the recognition that human life is characterized by confusion, ignorance, despair, and suffering. This recognition is reflected in the great ambitions of those ancient philosophical traditions to liberate human beings from suffering, and to edify and improve human life. Many forms of suffering were recognized in these ancient traditions, but central among them was physical illness—'sickness is suffering', says the First Noble Truth taught by the Buddha, as are 'pain, grief, and despair'—and many ancient philosophers employed medical metaphors to describe their practice.

We have largely forgotten the phronetic origins of philosophy and the fact that this practice, in the form of the pursuit of the good life, comes naturally to us. The spontaneous impulse to philosophize needs to be actively discouraged, as anyone who has taught children philosophy will know. This impulse is reawakened in some people during times of existential crises, such as those caused by illness, trauma, or grief. This positive yet unexpected reawakening of philosophical impulse needs to be articulated. It can, in conjunction with the insight that illness does not preclude well-being, comfort us in the knowledge that illness may disrupt, but will not destroy, one's striving towards the good life. The good life contains illumination and eclipse, bounty and challenge. This mix is essential for a human life as we know it. The tension and dialectic between the two sides of life sustains and nourishes philosophical reflection and promotes an existential stance.

Susan Sontag (1978) describes illness as 'the night-side of life', a different kingdom requiring a passport to enter it from the kingdom of the well. She describes the social fantasy in which death and mortality are denied, leading to a sense that the world of the ill is different to and separate from the world of the healthy. This is, of course, an illusion. We all partake in the kingdom of the ill, at least in principle. We all begin in complete dependence on others and we mostly end there.

Transience marks human life. The two kingdoms are connected and the transition from one to the other is the one certain fact of human life as it currently is. The world of the ill is different in many ways to the world of the healthy. Its space and time are different; its limitations obscure possibilities; it requires the ill to live in tense proximity to suffering and death. But the world of the ill is dependent upon the world of the healthy for its norms; and the world of the healthy is dependent on the world of the ill for the aberration of these norms. The two kingdoms mutually imply one another. We live in both, though we do much to deny this.

* * *

It is autumn and the park near our house is covered in dry leaves. The leaves are a warm glow of yellow and orange and brown, inviting us to roll in them, throw them in the air. My dog is running through big piles of leaves, excited at the autumn smells. I see a mother walking with her two sons. She bends down playfully and stuffs handfuls of leaves down their coats. They laugh and chase her round a tree, seeking retaliation. She runs, bending quickly to pick up more leaves, throwing them in the air, racing, chasing, and being chased. They laugh and call out to each other. So much air is needed for this, I think. I walk behind them, slowly, observing their playfulness, their physical joy. I share in their joy second-hand, but I am not able to join in anymore. I smile as I trudge up the hill, slowly, slowly, ever mindful of the oxygen so freely available in the air around us and so invisible to those who can have as much of it as they want.

Bibliography

Adorno, T. 1973. *The jargon of authenticity*, trans. K. Tarnowski and F. Will. London: Routledge and Kegan Paul.

Albrecht, G. L. and P. J. Devlieger. 1999. The disability paradox: high quality of life against all odds. *Social Science & Medicine* 48: 977–88.

Angner, E., M. N. Ray, K. G. Saag, and J. J. Allison. 2009. Health and happiness among older adults. *Journal of Health Psychology* 14(4): 503–12.

Annas, J. 2008. Happiness as achievement, in S. M. Cahn and C. Vitrano. (eds), *Happiness*. Oxford and New York: Oxford University Press, pp. 238–45.

Aurelius, Marcus. 1995. *Meditations*. New York and London: Penguin Books.

Barnes, P. J. and S. Kleinert. 2014. COPD—a neglected disease. *The Lancet* 364: 564–5.

Bauby, J.-D. 2007. *The diving bell and the butterfly*. London: Harper Perennial.

Beckman, H. and R. Frankel. 1984. The effect of physician behavior on the collection of data. *Annals of Internal Medicine* 101: 692–6.

Bernet, R. 2013. The body as a 'legitimate naturalization of consciousness'. *Philosophy* 72: 43–65.

Bernstein, J. M. 2011. Trust: on the real but almost always unnoticed, ever-changing foundation of ethical life. *Metaphilosophy* 42 (4): 395–416.

Biggs, J. and N. Boulton. 2011. *Voices from the shadows*. DVD documentary, UK.

Bishop, J. 2011. *The anticipatory corpse*. Notre Dame: Notre Dame University Press.

Blankenburg, W. 2002. First steps toward a psychopathology of 'common sense'. *Philosophy, Psychiatry and Psychology* 8(4): 303–15.

Blattner, W. 1994. The concept of death in *Heidegger's being and time*. *Man and World* 27: 49–70.

Blease, C. 2012. Stigmatising depression: folk theorising and 'the Pollyanna backlash', in H. Carel and R. Cooper (eds), *Health, illness and disease: philosophical essays*. Durham: Acumen, pp. 181–96.

Boethius. 2004. *The consolation of philosophy*, trans. H. R. James. <http://www.gutenberg.org/files/14328/14328-h/14328-h.htm> (accessed 20 August 2013).

Boorse, C. 1977. Health as a theoretical concept. *Philosophy of Science* 44(4): 542–73.

Bowden, H. 2012. A phenomenological study of anorexia nervosa. *Philosophy, Psychiatry and Psychology* 19(3): 227–41.

Brennan, J. 2012. Transitions in health and illness: realist and phenomenological accounts of adjustment to cancer, in H. Carel and R. Cooper (eds), *Health, illness and disease*. Durham: Acumen, pp. 129–42.

Brickman, P., D. Coates, and R. Janoff-Bulman. 1978. Lottery winners and accident victims: is happiness relative? *Journal of Personality and Social Psychology* 36(8): 917–27.

Buber, M. 2010 [1923]. *I and thou*. Eastford CT: Martino Publishing.

Burley, M. 2011. Emotion and anecdote in philosophical argument: the case of Havi Carel's *Illness*. *Metaphilosophy* 42(1–2): 33–48.

Bury, M. 1982. Chronic illness as biographical disruption. *Sociology of Health and Illness* 4(2): 167–82.

Cahn, S. M. and C. Vitrano. 2008. *Happiness*. Oxford and New York: Oxford University Press.

Calvo, P. and T. Gomila. 2008. *Handbook of cognitive science: an embodied approach*. Oxford: Elsevier.

Canguilhem, Georges. 1991. *The Normal and the pathological*. New York: Zone Books.

Carel, H. 2006. *Life and death in Freud and Heidegger*. New York: Rodopi.

Carel, H. 2007. Can I be ill and happy? *Philosophia* 35(2): 95–110.

Carel, H. 2009. 'I am well, apart from the fact that I have cancer': explaining wellbeing within illness, in L. Bortolotti (ed.), *Philosophy and happiness*. Basingstoke: Palgrave, pp. 82–99.

Carel, H. 2010. Phenomenology and its application in medicine. *Theoretical Medicine and Bioethics* 32(1): 33–46.

Carel, H. 2012. Phenomenology as a resource for patients. *Journal of Medicine and Philosophy* 37(2): 96–113. doi: 10.1093/jmp/JHS008.

Carel, H. 2013a. *Illness*. London: Routledge.

Carel, H. 2013b. Illness, phenomenology, and philosophical method. *Theoretical Medicine and Bioethics* 34(4): 345–57. doi: 10.1007/s11017-013-9265-1.

Carel, H. 2015. With bated breath: diagnosis of respiratory illness. *Perspectives in Biology and Medicine* 58(1): 53–65.

Carel, H. and I. J. Kidd. 2014. Epistemic injustice in healthcare: a philosophical analysis. *Medicine, Healthcare and Philosophy* 17(4): 529–40. DOI 10.1007/s11019-014-9560-2.

Carel, H. and J. Macnaughton. 2012. "How do you feel?": oscillating perspectives in the clinic. *The Lancet*: 2334–5. DOI:10.1016/S0140-6736(12)61007-1.

Carel, H., J. Macnaughton, and J. Dodd. 2015. The invisibility of breathlessness. *The Lancet Respiratory Medicine* 3(4): 278–9.

Carel, H. and D. Meacham. 2013 (eds). Human experience and nature. *Philosophy* 72, special supplement. Cambridge: Cambridge University Press.

Carman, T. 1994. On being social: a reply to Olafson. *Inquiry* 37: 203–23.

Carman, T. 1999. The body in Husserl and Merleau-Ponty. *Philosophical Topics* 27(2): 205–26.

Carman, T. 2000. Must we be inauthentic?, in M. Wrathall and J. Malpas (eds), *Heidegger, authenticity, and modernity*. Cambridge MA: MIT Press, pp. 13–28.

Carman, T. 2003. *Heidegger's analytic*. Cambridge: Cambridge University Press.

Carman, T. 2005. Authenticity, in H. Dreyfus and M. Wrathall (eds), *The Blackwell companion to Heidegger*. London: Blackwell, pp. 285–96.

Casado da Rocha, A. 2009. Towards a comprehensive concept of patient autonomy. *The American Journal of Bioethics* 9(2): 37–8.

Chanter, T. 2001. *Time, death and the feminine*. Stanford CA: Stanford University Press.

Chaung, H. T., G. M. Devins, J. Hunsley, and M. J. Gill. 1989. Psychosocial distress and wellbeing among gay and bisexual men with Human Immunodeficiency Virus infection. *American Journal of Psychiatry* 146(7): 876–80.

Choron, J. 1963. *Death and western thought*. New York: Collier Books.

Choron, J. 1972. *Death and modern man*. New York: Collier Books.

Chwalisz, K., E. Diener, and D. Gallagher. 1988. Autonomic arousal feedback and emotional experience: evidence from the spinal cord injured. *Journal of Personality and Social Psychology* 54(5): 820–8.

Cicero 1971. *On old age*, I Selected Works, trans. Michael Grant. London: Penguin.

Clark, A. 1997. *Being there*. Cambridge MA: MIT Press.

Clark, A. 2008. *Supersizing the mind*. Oxford: Oxford University Press.

Coady, D. 2010. Two concepts of epistemic injustice. *Episteme* 7(2): 101–13.

Coady, D. 2012. Critical reply to 'culpability for epistemic injustice: deontic or aretaic?', *Social Epistemology Review and Reply Collective* 1(5): 3–6.

Cottingham, J. (ed.) 2012. Special Issue: Aging and the elderly. *Philosophical Papers* 41(3).

Coulter, A. and J. Ellins. 2006. *Patient-focused interventions: a review of the evidence*. London: The Health Foundation. <http://www.health.org.uk> (accessed 4 May 2016).

Crichton, P., H. Carel and I. J. Kidd. Epistemic injustice and psychiatry. *BJPsych Bulletin* (in press).

Csordas, T. J. 2008. Intersubjectivity and intercorporeality. *Subjectivity* 22(1): 110–21.

Cutler, B. L. and S. D. Penrod. 1995. *Mistaken identifications: the eyewitness, psychology, and law*. New York: Cambridge University Press.

Dastur, F. 1996. *Death: an essay on finitude*, trans. John Llewelyn. London: Athlone Press.

Dennett, D. 1981. *Brainstorms*. Cambridge, MA: MIT Press.

Descartes, R. 1988 [1637]. Discourse on method, in *Descartes: selected philosophical writings*, trans. J. Cottingham, R. Stoothoff, and D. Murdoch. Cambridge: Cambridge University Press, pp. 20–56.

Descartes, R. 1988 [1641]. Meditations on first philosophy, in *Descartes: selected philosophical writings*, trans. J. Cottingham, R. Stoothoff, and D. Murdoch. Cambridge: Cambridge University Press, pp. 73–122.

Dolan, P. 1997. Modelling valuations for EuroQoL health states. *Medical Care* 35(11): 1095–108.

Dolezal, L. 2015. *The body and shame: phenomenology, feminism, and the socially shaped body*. New York: Rowman and Littlefield.

Dorfman, E. 2013. Naturalism, objectivism and everyday life, in Havi Carel and Darian Meacham (eds), *Human experience and nature: examining the relationship between phenomenology and naturalism*, vol. 72. Cambridge: Cambridge University Press, pp. 117–34.

Dorfman, E. 2014. *Foundations of the everyday: shock, deferral, repetition*. New York: Rowman and Littlfield.

Dreyfus, H. 1991. *Being-in-the-world: a commentary on Heidegger's* being and time, *Division I*. London and Cambridge MA: MIT Press.

Dreyfus, H. 2005. Foreword, in C. White, *Time and death*. London: Ashgate, pp. vi–xviii.

Drummond, John. 2007. *Historical dictionary of Husserl's philosophy*. Lanham MD: Scarecrow Press.

Earle, V. 2010. Phenomenology as research method or substantive metaphysics? An overview of phenomenology's uses in nursing. *Nursing Philosophy* 11(4): 286–96.

Edwards, P. 1975. Heidegger and death as 'possibility'. *Mind* 84: 548–66.

Edwards, P. 1976. Heidegger and death—a deflationary critique. *The Monist* 59(2): 161–86.

Edwards, P. 1979. *Heidegger on death: a critical evaluation*. La Salle: The Hegeler Institute.

Ehrenreich, B. 2010. *Smile or die: how positive thinking fooled America and the world*. London: Granta Books.

Emanuel, E. J. and L. L. Emanuel. 1992. Four models of the physician-patient relationship. *Journal of the American Medical Association*, April 22–267(16): 2221–6.

Epictetus. 2004. *A selection from the discourses of Epictetus with the Encheiridion*, trans. G. Long. <http://www.gutenberg.org/files/10661/10661-h/10661-h.htm> (accessed 20 August 2013).

Epicurus. 1994. *The Epicurus reader*. B. Inwood and L. P. Gerson (eds). Cambridge MA: Hackett.

Feldman, F. 2010. *What is this thing called happiness?* Oxford: Oxford University Press.

Feldman, M. D. 2004. *Playing sick? Untangling the web of Munchausen syndrome, Munchausen by proxy, malingering, and factitious disorder*. New York: Brunner-Routledge.

Frank, A. 1991. *At the will of the body*. New York: Mariner Books.

Frank, A. 2010. *The wounded storyteller: body, illness, and ethics*. Chicago: University of Chicago Press.

Freud, S. 1985 [1915]. Thoughts for the times on war and death, in the *Penguin Freud Library* vol. 12, *Civilization, society and religion*. London: Penguin Books.

Fricker, M. 2007. *Epistemic injustice: Power and the ethics of knowing*. Oxford: Oxford University Press.

Fulford, K.W., M. Broome, G. Stanghellini, and T. Thornton. 2005. Looking with both eyes open: fact and value in psychiatric diagnosis? *World Psychiatry* 4(2): 78–86.

Gallagher, S. 2005. *How the body shapes the mind*. Oxford: Oxford University Press.

Gallagher, S. 2009. Illness as a complete form of existence. Unpublished presentation delivered at the Association for Medical Humanities Annual Conference, Durham University.

Gallagher, S. and D. Zahavi. 2008. *The phenomenological mind*. New York: Routledge.

Gardner, S. 1999. *Kant and the* critique of pure reason. New York: Routledge.

Gawande, A. 2015. *Being mortal: illness, medicine and what matters in the end*. New York: Profile Books.

Gerhardt, S. 2004. *Why love matters: how affection shapes a baby's brain*. New York: Routledge.

Getz, L., A. L. Kirkengen, and E. Ulvestad. 2011. The human biology—saturated with experience. *Tidsskrift for Den Norske Legeforening* 7: 683–7.

Gilbert, D. 2006. *Stumbling on happiness*. London: Harper Press.

Goffman, E. 1963. *Stigma: notes on the management of spoiled identity*. New York: Simon and Schuster.

GOLD. 2010. Global strategy for the diagnosis, management, and prevention of chronic obstructive pulmonary disease (Updated). <http://www.ncbi.nlm.nih.gov> (accessed 4 May 2016).

Goldacre, B. 2009. *Bad science*. London: Harper Perennial.

Goldacre, B. 2012. *Bad pharma*. London: Fourth Estate.

Greener, I. 2009. *Healthcare in the UK: understanding continuity and change*. Bristol: The Policy Press.

Gysels, M. and I. J. Higginson. 2008. Access to services for patients with chronic obstructive pulmonary disease: the invisibility of breathlessness. *Journal of Pain and Symptom Management* 36 (5): 451–60.

Hadot, P. 1995. *Philosophy as a way of life*. London: Blackwell.

de Haes, J. C. J. M. and F. C. E. van Knippenberg. 1985. The quality of life of cancer patients: a review of the literature. *Social Science and Medicine* 20(8): 809–17.

de Haes, J. C. J. M., F. C. E. van Knippenberg, and J. P. Neijt. 1990. Measuring psychological and physical distress in cancer patients. *British Journal of Cancer* 62: 1034–8.

Haidt, J. 2006. *The happiness hypothesis*. London: William Heinemann.

Hass, L. 2008. *Merleau-Ponty's philosophy*. Bloomington: Indiana University Press.

Hatab, L. J. 1995. Ethics and finitude: Heideggerian contributions to moral philosophy. *International Philosophical Quarterly* 35(4): 403–17.

Hausman, D. 2014. *Valuing health*. Oxford: Oxford University Press.

Haybron, D. 2010. *The pursuit of unhappiness: the elusive psychology of well-being*. Oxford: Oxford University Press.

Heidegger, M. 1962 [1927]. *Being and time*. London: Blackwell.

Heidegger, M. 1984. *The metaphysical foundations of logic*, trans. M. Heim. Bloomington: Indiana University Press.

Heidegger, M. 1990. *Kant and the problem of metaphysics*, trans. R. Taft. Bloomington: Indiana University Press.

Heidegger, M. 1992. *History of the concept of time*, trans. T. Kisiel. Bloomington: Indiana University Press.

Heidegger, M. 1993. The origin of the work of art, in D. F. Krell (ed.), *Basic writings*. New York: Routledge, pp. 143–212.

Heidegger, M. 2000. *Introduction to metaphysics*, trans. G. Fried and R. Polt. London and New Haven CT: Yale University Press.

Helson, H. 1964. *Adaptation-level theory: an experimental and systematic approach to behaviour*. New York: Harper and Row.

Herigstad, M., A. Hayen, K. Wiech, and K. T. S. Pattinson. 2011. Dyspnoea and the brain. *Respiratory Medicine* 105(6): 809–17.

Heyes, C. 2012. Child, birth: an aesthetic, in L. Folkmarson Käll (ed.), *Dimensions of Pain*. London: Routledge, pp. 132–41.

Hofmann, B. 2002. On the triad disease, illness and sickness. *Journal of Medicine and Philosophy* 27(6): 651–73.

Hookway, C. 2010. Some varieties of epistemic injustice: response to Fricker. *Episteme* 7: 151–63.

Hope, L., W. Lewinski, J. Dixon, D. Blocksidge, and F. Gabbert. 2012. Witnesses in action: the effect of physical exertion on recall and recognition. *Psychological Science* 23(4): 386–90.

Hume, D. 1896 [1739]. *A treatise of human nature*. Oxford: Clarendon Press.

Hume, D. 1975 [1777]. *Enquiries concerning human understanding and concerning the principles of morals*. Oxford: Clarendon Press.

Hume, D. 2008 [1779]. *Dialogues concerning natural religion*. J. C. A. Gaskin (ed.). Oxford: Oxford Paperback.

Husserl, E. 1988 [1931]. *Cartesian meditations*. Dordrecht: Kluwer.

Husserl, E. 1989 [1952]. *Ideas pertaining to a pure phenomenology and to a phenomenological philosophy. Second book.* Dordrecht: Kluwer.

Husserl, E. 1990 [1928]. *On the phenomenology of the consciousness of internal time.* Dordrecht: Kluwer.

Husserl, E. 1997a [1948]. *Experience and judgement.* Evanston IL: Northwestern University Press.

Husserl, E. 1997b [1907]. *Thing and space: lectures of 1907.* Dordrecht: Kluwer.

Husserl, E. 2001 [1966]. *Analyses concerning passive and active synthesis: lectures on transcendental logic.* Dordrecht: Springer.

Innocence Project, The. <http://www.innocenceproject.org/causes-wrongful-con viction/eyewitness-misidentification> (accessed 26 October 2015).

Jaspers, K. 1997. *General psychopathology*, trans. J. Hoenig and Marian W. Hamilton. Baltimore and London: Johns Hopkins University Press.

Johnson, H. M. 2003. Unspeakable conversations. *New York Times* 16 February 2003.

Johnson, M. J., D. C. Currow, and S. Booth. 2014. Prevalence and assessment of breathlessness in the clinical setting. *Expert Reviews in Respiratory Medicine* 8(2): 151–61. doi: 10.1586/17476348.2014.879530.

Jones, P. W. 2001. Health status measurement in chronic obstructive pulmonary disease. *Thorax* 56: 880–7.

Kant, I. 1999 [1781]. *Critique of pure reason.* Cambridge: Cambridge University Press.

Keane, N. 2014. The hiddenness of health, in D. Meacham (ed.), *Medicine and society: new continental perspectives.* Dordrecht: Springer.

Kesserling, Amy. 1990. The experienced body, when taken-for-grantedness falters: a phenomenological study of living with breast cancer. PhD dissertation available via UMI.

Kidd, I. J. 2012. Can illness be edifying? *Inquiry* 55(5): 496–520.

Kidd, I. J. 2013. A pluralist challenge to 'integrative medicine': Feyerabend and Popper on the cognitive value of alternative medicine. *Studies in History and Philosophy of Biological and Biomedical Sciences* 44(3): 392–400.

Kidd, I. J. Transformative suffering and the cultivation of virtues. *Philosophy, Psychiatry, and Psychology* (forthcoming).

Kidd, I. J. and H. Carel. 2016. Epistemic injustice and illness. *Journal of Applied Philosophy* doi: 10.1111/japp.12172.

Kirkengen, A. L. 2007. Heavy burdens and complex disease—an integrated perspective. *Journal of the Norwegian Medical Association* 127: 3228–31.

Kleinman, A. 1980. *Patients and healers in the context of culture: an exploration of the borderland between anthropology, medicine, and psychiatry.* Berkeley CA: University of California Press.

Kleinman, A. 1988. *The illness narratives: suffering, healing, and the human condition.* New York: Basic Books.

Korsch, B. M., E. K. Gozzi, and V. Francis. 1968. Gaps in doctor–patient communication: doctor–patient interaction and patient satisfaction. *Pediatrics* 42: 855–71.

Korsch, B. M., E. K. Gozzi, and V. Francis. 1969. Gaps in doctor–patient communication ii: patients' response to medical advice. *The New England Journal of Medicine* 280: 535–40.

Krupp, A. 2009. *Reason's children: childhood in early modern philosophy*. Lewisburg PA: Bucknell University Press.

Lakoff, G. and M. Johnson. 1999. *Philosophy in the flesh*. New York: Basic Books.

Law, I. 2009. Motivation, depression and character, in M. R. Broome and L. Bortolotti (eds), *Psychiatry as cognitive neuroscience: philosophical perspectives*. Oxford: Oxford University Press, pp. 351–64.

Leder, D. 1990. *The absent body*. Chicago: University of Chicago Press.

Leman-Stefanovic, I. 1987. *The event of death: a phenomenological inquiry*. Dordrecht: Martinus Nijhoff.

Leontiou, J. F. 2010. *What do the doctors say?* New York: iUniverse.

Levinas, E. 1969 [1961]. *Totality and infinity: an essay on exteriority*, trans. A. Lingis. Pittsburgh PA: Duquesne University Press.

Levinas, E. 1998. 'Dying for . . . ', in E. Levinas (ed.), *Entre nous: on thinking-of-the-other*, trans. M. B. Smith and B. Harshaw. London: Athlone Press, pp. 179–88.

Lewis, M. 2005. *Heidegger and the place of ethics*. London: Continuum.

Lindqvist, O., A. Widmark, and B. Rasmussen. 2006. Reclaiming wellness—living with bodily problems as narrated by men with advanced prostate cancer. *Cancer Nursing* 29(4): 327–37.

Lindsey, E. 1996. Health within illness: experiences of chronically ill/disabled people. *Journal of Advanced Nursing* 24: 465–72.

Little, M., C. Jordens, K. Paul, K. Montgomery, and B. Philipson. 1998. Liminality: a major category of the experience of cancer illness. *Social Science & Medicine* 47(10): 1485–94.

Lloyd, G. 1984. *The man of reason*. London: Routledge.

Lykken, D. and A. Tellegen. 1996. Happiness is a stochastic phenomenon. *Psychological Science* 7(3): 186–9.

Lyubomirsky, S. 2007. *The how of happiness*. London: Piatkus Books.

Macann, C. 1992. *Critical assessments*. New York and London: Routledge.

MacIntyre, A. 1999. *Dependent rational animals*. London: Duckworth.

MacKinnon, C. 1993. *Only words*. Boston: Harvard University Press.

Mansbach, A. 1991. Heidegger on the self, authenticity and inauthenticity. *Iyun* 40: 65–91.

Marotta, J. J. and M. Behrmann. 2004. Patient Schn: has Goldstein and Gelb's case withstood the test of time? *Neuropsychologia* 42: 633–8.

McIver, S. 2011. User perspectives and involvement, in K. Walshe and J. Smith (eds), *Healthcare management* 2nd ed. Maidenhead: Open University Press, pp. 354–72.

Meacham, D. 2013. What goes without saying: Husserl's notion of style. *Research in Phenomenology* 43: 3–26.

Medina, J. 2012. *The epistemology of resistance: gender and racial oppression, epistemic injustice, and the social imagination.* Oxford: Oxford University Press.

Merleau-Ponty, M. 1964a. Cezanne's doubt, in M. Merleau-Ponty, *Sense and nonsense,* trans. P. A. Dreyfus and H. L. Dreyfus. Evanston IL: Northwestern University Press, pp. 9–25.

Merleau-Ponty, M. 1964b. *The primacy of perception,* trans. W. Cobb. Evanston IL: Northwestern University Press.

Merleau-Ponty, M. 2012 [1945]. *Phenomenology of perception.* New York: Routledge.

Michael, S. R. 1996. Integrating chronic illness into one's life. *Journal of Holistic Nursing* 14(3): 251–67.

Mill, J. S. 1989 [1873]. *Autobiography.* London: Penguin Classics.

Montaigne, M. de. 1993 [1580]. To philosophize is to learn how to die, in M. A. Screech (ed.), *The essays: a selection.* London: Penguin, pp. 17–36.

Moran, D. 2000. *Introduction to phenomenology.* London: Routledge.

Moran, D. 2010. Husserl, Sartre and Merleau-Ponty on embodiment, touch, and the 'double sensation', in K. Morris (ed.), *Sartre on the body.* London: Palgrave Macmillan, pp. 41–66.

Mulhall, S. 2005. Human mortality: Heidegger on how to portray the impossible possibility of Dasein, in H. Dreyfus and M. Wrathall (eds), *The Blackwell companion to Heidegger.* London: Blackwell, pp. 297–310.

Murphy, D. 2006. *Psychiatry in the scientific image.* Cambridge MA: MIT Press.

Nabokov, V. 1959 [1935]. *Invitation to a beheading,* trans. D. Nabokov. London: Weidenfeld and Nicholson.

Nancy, J. L. 1993. *The birth to presence.* Stanford: Stanford University Press.

National Institute for Health and Clinical Excellence (NICE). Depression in adults with a chronic physical health problem: treatment and management. 2009. Clinical guideline 91. <http://www.nice.org.uk/CG91> (accessed 8 March 2016).

NHS Constitution. 2013. <http://www.nhs.uk/choiceintheNHS/Rightsandpledges/NHSConstitution/Documents/2013/the-nhs-constitution-for-england-2013.pdf> (accessed 11 August 2014).

Nietzsche, Friedrich. 2004. Why I am so wise, in F. Nietzsche, *Ecce homo.* Oxford: Oxford University Press, pp. 7–18.

Noë, A. 2004. *Action in perception*. London: MIT Press.

Nussbaum, M. 1990. *Love's knowledge*. Oxford: Oxford University Press.

Nussbaum, M. 1994. *The therapy of desire: theory and practice in Hellenistic ethics*. Princeton NJ: Princeton University Press.

Okasha, S. 2003. Probabilistic induction and Hume's problem: reply to Lange. *Philosophical Quarterly* 53(212): 419–24.

Øverenget, E. 1998. *Seeing the self*. Dordrecht: Kluwer Academic Publishers.

Parsons, T. 1971. *The system of modern societies*. New Jersey: Prentice-Hall.

Paterson, B. 2001. The shifting perspectives model of chronic illness. *Journal of Nursing Scholarship* 33(1): 21–6.

Patients Association. <http://www.patients-association.org.uk/about/> (accessed 11 August 2014).

Paul, L. A. 2014. *Transformative experience*. Oxford: Oxford University Press.

Pausch, R. 2008. *The last lecture*. London: Hodder and Stoughton.

Petitot, J., F. Varela, B. Pachoud, and J. M. Roy. 1999. *Naturalizing phenomenology*. Stanford: Stanford University Press.

Pezdek, K., S. T. Lam, and K. Sperry. 2009. Forced confabulation more strongly influences event memory if suggestions are other-generated than self-generated. *Legal and Criminological Psychology* 14: 241–52.

Philipse, H. 1998. *Heidegger's philosophy of being*. Princeton NJ: Princeton University Press.

Plumwood, V. 1993. *Feminism and the mastery of nature*. London: Routledge.

Polt, R. 1999. *Heidegger: an introduction*. New York and London: Routledge.

Porter, R. 1999. *The greatest benefit to mankind*. New York: Fontana Press.

Ratcliffe, M. 2008a. *Feelings of being: phenomenology, psychiatry and the sense of reality*. Oxford: Oxford University Press.

Ratcliffe, M. 2008b. Touch and situatedness. *International Journal of Philosophical Studies* 16(3): 99–322.

Ratcliffe, M. 2011. What is it to lose hope? *Phenomenology and Cognitive Science* 12: 597–614. doi: 10.1007/s11097-011-9215-1.

Ratcliffe, M. 2012a. Varieties of temporal experience in depression. *Journal of Medicine and Philosophy* 37(2): 114–38.

Ratcliffe, M. 2012b. Phenomenology as a form of empathy. *Inquiry* 55(5). doi:10.1080/0020174X.2012.716196.

Ratcliffe, M. 2013. Phenomenology, naturalism and the sense of reality. *Philosophy* 72: 67–88.

Ratcliffe, M., M. Broome, B. Smith, and H. Bowden. 2013. A bad case of the flu? The comparative phenomenology of depression and somatic illness. *Journal of Consciousness Studies* 20(7–8): 198–218.

Riggs, W. 2012. Culpability for epistemic injustice: deontic or aretaic? *Social Epistemology* 26(2): 149–62.

Riis, J., J. Baron, G. Loewenstein, and C. Jepson. 2005. Ignorance of hedonic adaptation to haemodialysis: a study using ecological momentary assessment. *Journal of Experimental Psychology* 134(1): 3–9.

Rodley, C. (ed.) 1992. *Cronenberg on Cronenberg.* London: Faber and Faber.

Rosenbaum, L. 2012. How much would you give to save a dying bird? Patient advocacy and biomedical research. *New England Journal of Medicine* 367: 1755–9. <http://www.nejm.org/doi/full/10.1056/NEJMms1207114> (accessed 19 March 2016).

Saks, E. 2007. *The center cannot hold: my journey through madness.* New York: Hyperion.

Sartre, J.-P. 2003 [1943]. *Being and nothingness.* London and New York: Routledge.

Savino, A. C. and J. S. Fordtran. 2006. Factitious disease: clinical lessons from case studies at Baylor University Medical Center. *Proceedings of Baylor University Medical Center* 19: 195–208.

Sayeau, M. 2009. How should we live? *Philosophers' Magazine* 44: 101–3.

Scarry, E. 1985. *The body in pain.* Oxford: Oxford University Press.

Schkade, D. A. and D. Kahneman. 1998. Does living in California make people happy? A focusing illusion in judgments of life satisfaction. *Psychological Science* 9(5): 340–6.

Schneider, C. E. 1998. *The practice of autonomy.* New York: Oxford University Press.

Scrutton, A. Two Christian theologies of depression: an evaluation and discussion of clinical implications. *Philosophy, Psychiatry, & Psychology* (forthcoming).

Seneca. 2004. *On the shortness of life.* London: Penguin Books.

Shakespeare, W. 2007. *As you like it,* in *The RSC Shakespeare: the complete works.* Basingstoke: Palgrave Macmillan.

Smith, A. D. 2003. *Husserl and the* Cartesian Meditations. London and New York: Routledge.

Smith, D. W. 2007. *Husserl.* London and New York: Routledge.

Sontag, S. 1978. *Illness as metaphor.* New York: Farrar, Straus and Giroux.

Stambaugh, J. 1978. An inquiry into authenticity and inauthenticity, in J. Sallis (ed.), Being and time: *radical phenomenology* essays in honor of Martin Heidegger. New Jersey: Humanities Press, pp. 153–61.

Stanghellini, G. 2004. *Disembodied spirits and deanimated bodies.* Oxford: Oxford University Press.

Svenaeus, F. 2000a. Das Unheimliche—towards a phenomenology of illness. *Medicine, Health Care and Philosophy* 3: 3–16.

Svenaeus, F. 2000b. The body uncanny—further steps towards a phenomenology of illness. *Medicine, Health Care and Philosophy* 3: 125–37.

Svenaeus, F. 2001. *The hermeneutics of medicine and the phenomenology of health*. Linköping: Springer.

Svenaeus, F. 2012. Organ transplantation and personal identity: how does the loss and change of organs have effects on the self? *Journal of Medicine and Philosophy* 37(2): 139–58. doi: 10.1093/jmp/jhs011.

Sweet, V. 2006. *Rooted in the earth, rooted in the sky: Hildegard of Bingen and premodern medicine*. New York: Routledge.

Taylor, S. E., R. L. Falke, S. J. Shoptaw, and R. R. Lichtman. 1986. Social support, support groups, and the cancer patient. *Journal of Consulting and Clinical Psychology* 54(5): 608–15.

Thomas, D. 1995. *The Dylan Thomas omnibus: under milk wood, poems, stories, broadcasts*. London: J. M. Dent.

Thorne, S. and B. Paterson. 1998. Shifting images of chronic illness. *Image: Journal of Nursing Scholarship* 30(2): 173–8.

Thorne, S., B. Paterson, S. Acorn, C. Canam, G. Joachim, and C. Jillings. 2002. Chronic illness experience: insights from a metastudy. *Qualitative Health Research* 12(4): 437–52.

Thorsrud, H. 2009. *Ancient scepticism*. Stocksfield: Acumen.

Tiberius, V. 2010. *The reflective life: living wisely with our limits*. Oxford: Oxford University Press.

Tolstoy, L. 1995 [1886]. *The death of Ivan Ilyich*. London: Penguin Books.

Toombs, S. K. 1987. The meaning of illness: a phenomenological approach to the patient–physician relationship. *Journal of Medicine and Philosophy* 12: 219–40.

Toombs, S. K. 1988. Illness and the paradigm of lived body. *Theoretical Medicine* 9: 201–26.

Toombs, S. K. 1990. The temporality of illness: four levels of experience. *Theoretical Medicine* 11: 227–41.

Toombs, S. K. 1993. *The meaning of illness: a phenomenological account of the different perspectives of physician and patient*. Amsterdam: Kluwer.

Toombs, S. K. 1995. The lived experience of disability. *Human Studies* 18: 9–23.

Toombs, S. K. 2001. The role of empathy in clinical practice. *Journal of Consciousness Studies* 8(5–7): 247–58.

Twaddle, A. 1968. Influence and illness. PhD thesis, Brown University.

Ubel, P. A., G. Loewenstein, N. Schwarz, and D. Smith. 2005. Misimagining the unimaginable. *Health Psychology* 24 (4): 57–62.

Varelius, J. 2006. The value of autonomy in medical ethics. *Medicine, Health Care and Philosophy* 9(3): 377–88.

Visker, R. 1996. Dropping—the 'subject' of authenticity: *being and time* on disappearing existentials and true friendship with being, in S. Critchley and P. Dews (eds), *Deconstructive subjectivities*. Albany NY: State University of New York Press, pp. 59–83.

Vogel, L. 1994. *The fragile 'we'*. Evanston: Northwestern University Press.

Vogt, K. 2014. *Stanford encyclopaedia of philosophy* <http://plato.stanford.edu/entries/skepticism-ancient/> (accessed 28 July 2014).

Wainwright, M. and J. Macnaughton. 2013. Is a qualitative perspective missing from COPD guidelines? *The Lancet Respiratory Medicine* 1(6): 441–2.

Wakefield, J. 1992. Disorder as harmful dysfunction: a conceptual critique of DSM-III-R's definition of mental disorder. *Psychological Review* 99(2): 232–47.

Ware, B. 2012. *The top five regrets of the dying: a life transformed by the dearly departing*. London: Hay House UK.

Weinstein, N. 1980. Unrealistic optimism about future life events. *Journal of Personality and Social Psychology* 39: 806–20.

Wheeler, M. 2005. *Reconstructing the cognitive world*. Cambridge MA: MIT Press.

Wheeler, M. 2013. Science friction: phenomenology, naturalism and cognitive science, in H. Carel and D. Meacham (eds), *Human experience and nature: examining the relationship between phenomenology and naturalism*. vol. 72. Cambridge: Cambridge University Press, pp. 135–68.

Wiggins, O. P. and J. Z. Sadler (eds) 2005. Clinical ethics of Richard M. Zaner. Special issue. *Theoretical Medicine and Bioethics* 26(1): 1–104.

Williams, S. J. 2003. *Medicine and the body*. London: Sage Publications.

Withers, C. 2015. Illness as authenticity: applying Heideggerian phenomenology to the experience of illness. Unpublished MA dissertation, Bristol University.

Wittgenstein, L. 1974. *On certainty*. Oxford: Basil Blackwell.

Wootton, D. 2006. *Bad medicine*. Oxford: Oxford University Press.

Yoeli-Tlalim, R. 2010. Tibetan 'wind' and 'wind' illnesses: towards a multicultural approach to health and illness. *Studies in History and Philosophy of Biological and Biomedical Sciences* 41(4): 318–24.

Young, I. M. 2005a. Throwing like a girl: a phenomenology of feminine body comportment, motility and spatiality, in I. M. Young, *On female body experience: 'throwing like a girl' and other essays*. Oxford: Oxford University Press, pp. 27–45.

Young, I. M. 2005b. Pregnant embodiment: subjectivity and alienation, in I. M. Young, *On female body experience: 'throwing like a girl' and other essays*. Oxford: Oxford University Press, pp. 46–61.

Zahavi, D. 2003. *Husserl's phenomenology*. Stanford CA: Stanford University Press.

Zahavi, D. 2013. Naturalized phenomenology: a desideratum or a category mistake?, in H. Carel and D. Meacham (eds), *Human experience and nature: examining the relationship between phenomenology and naturalism*, vol. 72. Cambridge: Cambridge University Press, pp. 1–22.

Zaner, R. M. 1981. *The context of self*. Athens OH: Ohio University Press.

Zaner, R. M. 2005. A work in progress. *Theoretical Medicine* 26: 89–104.

Index

Printed and bound by CPI Group (UK) Ltd, Croydon, CR0 4YY